Diseases
of
Nematodes

Volume II

Editors

George O. Poinar, Jr.
Department of Entomological Sciences
University of California
Berkeley, California

Hans-Börje Jansson
Department of Microbial Ecology
Lund University
Lund, Sweden

CRC Press
Taylor & Francis Group
Boca Raton London New York

CRC Press is an imprint of the
Taylor & Francis Group, an **informa** business

First published 1988 by CRC Press
Taylor & Francis Group
6000 Broken Sound Parkway NW, Suite 300
Boca Raton, FL 33487-2742

Reissued 2018 by CRC Press

Library of Congress Cataloging-in-Publication Data

Diseases of Nematodes.

    Includes bibliographies and indexes.
    1. Nematoda — Diseases. Nematoda — Biological control.   I. Poinar, George O.   II. Jansson, Hans-Börje,
1947-   . [DNLM: 1. Nematoda 8 parasitology.   2. Pest Control, Biological.   QX 203 D611]
SF997.5.N44D57   1988   632'.65182   87-22435
ISBN 0-8493-4317-8 (v. 1)
ISBN 0-8493-4318-6 (v.2)

A Library of Congress record exists under LC control number: 87022435

Publisher's Note
The publisher has gone to great lengths to ensure the quality of this reprint but points out that some imperfections in the original copies may be apparent.

Disclaimer
The publisher has made every effort to trace copyright holders and welcomes correspondence from those they have been unable to contact.

ISBN 13: 978-1-315-89237-5 (hbk)
ISBN 13: 978-1-351-07147-5 (ebk)

Visit the Taylor & Francis Web site at http://www.taylorandfrancis.com and the
CRC Press Web site at http://www.crcpress.com

INTRODUCTION

The period of time nematodes have existed is not known due to the paucity of fossil records, especially Pre-Tertiary. However, some investigators place the phylum Nematoda as arising during the Cambrian or Pre-Cambrian periods. This would allow at least half a billion years for symbiotic associations to become established between nematodes and other agents in the environment. There is fossil evidence indicating that nematophagous fungi were present some 20 to 30 million years ago.

The present work deals with the diseases of nematodes. Although the term disease implies a pathological condition brought about by an infectious agent, a broader concept is used here. Disease is considered to be a departure from the state of health or normality in which the body, or a cell, tissue, or organ of the body, is disturbed functionally or structurally. This departure can manifest itself as a disruption of the growth, development, function, or adjustment of the organism to its environment. Thus, any agent which either by its presence or absence, causes any destructive process at the cellular or organismic level is a disease-producing agent.

Diseases can be considered infectious or noninfectious. Noninfectious diseases are covered in the first section of this work and arise from aberrations in the genetics, nutrition, metabolism, or physiology of the organism. Although the direct cause of noninfectious diseases are usually abiotic factors, such conditions are often indirectly related to the presence or absence of other organisms and their products (e.g., toxins produced by plants, starvation resulting from the absence of bacteria, injury through the action of predators).

Infectious diseases are caused by parasites or pathogens which develop at the expense of the host. Some parasites which live in intestinal lumen of nematodes (e.g., protozoa) may not cause any noticeable disturbance and therefore would not be considered pathogens (microorganisms capable of producing disease under normal conditions). The known pathogens of nematodes, which are treated in this work, are viruses, bacteria, protozoa, and rickettsia. Although fungi are considered pathogens in relation to insect disease, in the field of nematology they are generally grouped under parasites and predators, depending on their mode of infection and ability to exist saprophytically in the nematode's environment. The complete range of fungal parasites and predators of nematodes are treated here.

A chapter on the invertebrate predators of nematodes is also included, although this heading would not normally fall under disease. However, aside from its importance in relation to nematode population dynamics, invertebrate predation is often initiated through the action of toxins which have an instantaneous effect on the nematode or its tissues.

In some respects, the field of nematode diseases can be considered in its infancy, especially in areas involving viral, protozoal, and bacterial associations. However, it is clear that basic research will reveal many new and interesting types of pathogens and relationships, and it is hoped that this work will stimulate investigations along these lines. The evaluation of diseases in the population dynamics and distribution of nematode species is also worth attention.

Another purpose in creating this work was to focus attention on organisms which have potential for the biological control of nematodes. Millions of dollars are spent each year treating nematode pests of plants and animals. The use of nematicides for the control of plant parasitic nematodes has become a well established practice. However, with new legislation resulting from the potential dangers of these chemicals, the use of nematicides is being restricted, and alternative methods of control are becoming a necessity. Thus the concept of biological control of nematodes has now come out of the laboratory into the field arena. Investigators are attempting to make up for what can be considered a considerable lack of foresight some 20 to 30 years ago, when scientists interested in the diseases of nematodes were given little support.

EDITORS

George O. Poinar, Jr., Ph.D., is an invertebrate pathologist and nematologist currently on the faculty in the Department of Entomological Sciences at the University of California in Berkeley, California. He received his Ph.D. from Cornell University in 1962 with majors in Entomology, Nematology, and Botany.

After graduating, he studied nematology with J. B. Goodey in England, Oostenbrink in the Netherlands, Kirjanova and Rubtsov in the Soviet Union, and Chabaud in Paris. He also conducted nematode investigations in Western Samoa, New Caledonia, New Guinea, the Philippines, Malaysia, Thailand, and West Africa.

He has authored three books on nematodes and coauthored two books on insect pathogens and parasites. He has also published over 230 research papers in the areas of nematology, entomology, invertebrate pathology, biological control, and paleosymbiosis.

Hans-Börje Jansson, Ph.D., is Assistant Professor at the Department of Microbial Ecology, Lund University, Lund, Sweden, and a Researcher at the Swedish Natural Science Research Council.

Dr. Jansson graduated with a B.Sc. in biology and chemistry in 1972 from Lund University and received his Ph.D. degree in microbiology from the same university in 1982. During 1983 to 1985 Dr. Jansson visited the Department of Plant Pathology, University of Massachusetts, Amherst, Massachusetts as a Fulbright Fellow and Adjunct Visiting Professor.

Dr. Jansson has been studying several aspects of interactions between nematodes and nematophagous fungi and has published more than 40 research papers on this subject. His current research interest is in mechanisms involved in chemoattraction of fungal and nematode systems.

CONTRIBUTORS, VOLUME I

Jaap Bakker, Dr. Ir.
Associate Professor
Department of Nematology
Agricultural University
Wageningen, Netherlands

Robert S. Edgar, Ph.D.
Professor
Department of Biology
University of California at Santa Cruz
Santa Cruz, California

F. J. Gommers, Dr. Ir.
Associate Professor
Department of Nematology
Agricultural University
Wageningen, Netherlands

Eder L. Hansen, Ph.D.
Retired
Berkeley, California

James W. Hansen, Ph.D.
Retired
Berkeley, California

Roberta Hess
Department of Entomological Sciences
University of California at Berkeley
Berkeley, California

George O. Poinar, Jr., Ph. D.
Professor
Department of Entomological Sciences
University of California at Berkeley
Berkeley, California

Richard Martin Sayre, Ph. D.
Research Plant Pathologist
Nematology Laboratory
Agricultural Research Service
U.S. Department of Agriculture
Beltsville, Maryland

Mortimer P. Starr, Ph.D.
Professor Emeritus
Department of Bacteriology
University of California at Davis
Davis, California

CONTRIBUTORS, VOLUME II

N. F. Gray, Ph.D.
Lecturer in Environmental Science and
Fellow of Trinity College
Department of Environmental Science
Trinity College
University Of Dublin
Dublin, Ireland

Hans-Börje Jansson, Ph. D.
Assistant Professor
Department of Microbial Ecology
Lund University
Lund, Sweden

Gareth Morgan-Jones, Ph.D., D.Sc.
Professor
Department of Plant Pathology
Auburn University
Auburn, Alabama

Birgit Nordbring-Hertz, Ph.D.
Professor
Department of Microbial Ecology
Lund University
Lund, Sweden

Rodrigo Rodríguez-Kábana, Ph.D.
Professor
Department of Plant Pathology
Auburn University
Auburn, Alabama

Richard W. Small, Ph.D., MI Biol.
Department of Biology
Liverpool Polytechnic
Liverpool, England

Graham R. Stirling, Ph.D.
Senior Nematologist
Plant Pathology Branch
Queensland Department of Primary
 Industries
Indooroopilly, Queensland
Australia

The present work is dedicated to the memory of Robert Ph. Dollfus of the Muséum National d'Histoire Naturelle, Paris. Included among his many interests were the natural enemies of parasitic worms. His own observations, as well as those from the literature, were published in 1946 in the classic ''Parasites des Helminthes'', the first time this novel subject was consolidated into a single volume.

TABLE OF CONTENTS, VOLUME I

TABLE OF CONTENTS, VOLUME II

PARASITES AND PREDATORS

PARASITES AND PREDATORS

Chapter 1

FUNGI ATTACKING VERMIFORM NEMATODES

N. F. Gray

TABLE OF CONTENTS

I. INTRODUCTION

A wide and diverse range of fungi which feed on nematodes occur in the soil. This is not surprising when one considers the long co-evolution of nematodes and soil fungi which has inevitably occurred in the close confines of the soil habitat. So predaceous and parasitic relationships have evolved amongst most of the major groups of soil fungi from the Phycomycetes to the Basidomycetes.[1] Collectively they are known as nematophagous or nematode-destroying fungi, with the term nematode-trapping fungi used to describe predators only. They are natural enemies of nematodes and have developed very sophisticated strategies for either infecting or capturing these small animals. Nematophagous fungi fall into two broad groups, those which parasitize nematodes (endoparasites) with small conidia or zoospores, or those which capture nematodes (predators) using modified hyphal traps. The fate of the nematode will be the same whether infected by an endoparasite or captured by a predator, with fungal hyphae developing within the nematode and the body contents utilized by the fungus. In nature, nematophagous fungi help to recycle carbon, nitrogen, and other important elements from the often considerable biomass of soil nematodes which are feeding on the microbial decomposers. Due to the frequency at which nematophagous fungi are isolated, it is tempting to surmise that their role of nutrient recycling in the soil is a major one, although it has never been quantified. For man however, the ability of the group to capture and destroy nematodes has presented the attractive possibility of harnessing the fungi as a biological control for this most damaging agricultural pest.

The first nematophagous fungi to be isolated and described was *Arthrobotrys oligospora* by Fresenius in 1852.[2] He was unaware of its predatory habit which was first observed 36 years later by Zopf.[3] So the study of nematophagous fungi is now almost a century old. However, virtually nothing was known of these fungi until Charles Drechsler, working in Beltsville, Maryland, began his systematic study of the group. His vast contribution to our knowledge of these fungi, especially in the isolation and detailed description of new species between 1933 to 1975, and the work of others most notably Duddington in London (1950 to 1972), has provided the solid basis for research enjoyed today. At present over 150 species of nematophagous fungi are known, with new species reported each year as isolation techniques improve and new habitats are examined. The group has been extensively reviewed with a beautifully illustrated monograph produced by Barron.[4] More recent reviews have been prepared by Mankau,[1] Barron,[5,6] Peloille,[7] and Lysek and Nordbring-Hertz.[8]

II. THE NEMATOPHAGOUS HABIT

Nematophagous fungi can be categorized as either endoparasitic or predatory in habit. Endoparasites do not produce extensive mycelium but exist in the environment as conidia, which infect nematodes by either adhering to the surface of the prey or being ingested. The conidia rapidly germinate and invade the entire nematode with assimilative hyphae absorbing all the body contents. The life cycle is completed by the fungus breaking through the body wall of the host to form reproductive structures such as conidiophores which support conidia, or evacuation tubes from which zoospores are released. In contrast, predatory fungi produce extensive hyphal systems in the environment and produce trapping devices at intervals along each hypha. The traps capture nematodes either mechanically or by adhesion, allowing the fungus to rapidly penetrate the prey and digest the body contents.

The classification of species as either endozoic or predatory is usually quite straight forward, however some species lie between the two modes. For example, some species produce free-swimming, flagellate spores (zoospores) which are probably chemically attracted to nematodes by body excretions. The zoospores swim towards the nematodes and encyst on to their surface then penetrate and kill the nematode as a normal endoparasite.

Because such species actively pursue, catch, and kill nematodes, Barron[4] has suggested that they could be classed as predators, but in terms of morphology and life-cycle they are included with the endoparasites. The genus *Nematoctonus* forms a bridge between the two groups with some species predatory and others endoparasitic in habit. The predatory species capture nematodes by means of adhesive knobs growing directly from the hyphae, while adhesive knobs are never found on the hyphae of the endoparasitic species of the genus, rather nematodes are captured by adhesive knobs that form on the conidia only.

A. Endoparasites

In the soil, endozoic parasites of nematodes exist mainly as conidia. Most endoparasitic nematophagous fungi are obligate parasites, spending their entire vegetative lives within the infected host. There is no extensive hyphal development from infected nematodes, with only evacuation tubes, or conidiophores and conidia, produced externally. Infection is achieved by three distinct strategies. In the higher fungi, conidia are passive and either ingested, lodging in the digestive tract where they subsequently germinate, or are adhesive, and become attached to the surface of the nematode and penetrate through its cuticle. Some species of the lower fungi are more aggressive as parasites and produce flagellated spores which are attracted to potential hosts and encyst directly on to their cuticle. In general terms, the conidia of endoparasites are an order of magnitude smaller than the conidia of predators, with adhesive spores usually <4 μm in diameter. The reason for this is the conidia of endoparasites do not require large food resources in order to produce extensive vegetative hyphae, just enough energy to allow them to germinate and penetrate the host.

Endoparasites are found in a number of taxonomic classes which can be loosely categorized into three groups: Group I, Encysting species of the Chytridiomycetes (*Catenaria*) and Oomycetes (*Myzocytium*); Group II, Deuteromycetes producing adhesive conidia (*Verticillium, Cephalosporium, Meria*) and ingested conidia (*Harposporium*); and Group III, Basidiomycetes forming adhesive conidia (*Nematoctonus*). In a survey of Irish nematophagous fungi Gray[9] measured the frequency of occurrence of endoparasites isolated from a range of soils. He found that encysting (Group I) species represented 21.4% of isolations, Deuteromycetes (Group II) 67.1%, and Basidiomycetes (Group III) 11.4%. Adhesive Deuteromycetes were more frequently isolated (45.7%) than ingested conidia (21.4%).

1. Encysting Spores

The endoparasitic species of Chytridiomycetes and Oomycetes have similar methods of infecting nematodes, with flagellated zoospores produced in hyphal bodies (zoosporangia) and released into the environment via evacuation tubes. The zoospores have only a short time in which to find a suitable host and do this by following the chemical gradient formed by exudates released from the body orifices of nematodes. On reaching the nematode the zoospore encysts close to an orifice, i.e., the anus, vulva, or buccal cavity. Infection is achieved either by germ tubes which pass into the nematode through an orifice or by direct penetration through the animal's cuticle. Inside the host an infection thallus is formed which absorbs the body contents, and on reaching maturity sections of the thallus swell to form discrete zoosporangia in which zoospores are produced. The life cycle is completed when the zoospores are released into the soil environment via the evacuation tubes which connect each zoosporangium to outside the host's body. While this method is used by *Catenaria anguillulae* and many species of *Myzocytium*, other species have developed more complex methods of infection.[10,11]

2. Adhesive Conidia

Endoparasites forming adhesive conidia are found in a number of genera of the Deuteromycetes, most notably *Verticillium, Cephalosporium,* and *Meria*. The conidia are small and streamline in shape, and when mature they produce a strong adhesive bud at the distal end.[12] Ultrastructural studies on the conidia of *Meria coniospora* have shown that the apical knob is always covered with adhesive material in which radiating fibrils can be seen, and that attachment to the nematode is by means of this fibrillose layer.[13] Nematodes are chemically attracted to the conidia and on contact the conidium adheres to the surface of the host.[14] Although the conidia can adhere to any part of the nematode body they are mainly found on the head region or close to the nematode's active chemosensory structures.[15,16] The conidia are very difficult to dislodge and infected nematodes are regularly observed trying to remove such conidia by rubbing the infected body area on soil debris. Once attached, a narrow germ tube grows from the adhesive bud which penetrates the cuticle and forms a tiny infection bladder in the integument, from which assimilative hyphae rapidly grow to fill the body cavity. When the body contents are completely digested, which normally takes a few days, hyphae break through the nematode cuticle. These hyphae produce conidiophores which generally have short, slim lateral branches (phialides) on which conidia are formed. The attraction and infection of nematodes by *Meria coniospora* and *Cephalosporium balanoides,* two very common endoparasites, have been extensively examined.[14,15,17]

The endoparasitic Basidiomycetes also infect nematodes with adhesive conidia, but the conidia form adhesive knobs similar to those formed by the predatory species. Endoparasitic species are only separated from predatory species by the fact that no organs of capture are formed on the actual mycelium, only on the conidia. The hyphae of Basidiomycetes have characteristic clamp connections which are seen in both the predatory and endoparasitic species. Both the conidiophores and the assimilative hyphae of endoparasites bear these clamp connections. The life cycle is the same as for other endoparasites with adhesive conidia. Endoparasitic Basidiomycetes are widely distributed with *Nematoctonus leiosporus,* the most frequently isolated species.[4,9]

3. Ingested Conidia

Some species of endoparasites have developed morphologically adapted conidia, which when eaten by the host, become lodged in either its buccal cavity or oesophagus. These species belong almost exclusively to the genus *Harposporium,* and the degree of adaptation is quite unusual.[18,19] The conidia are crescent-shaped (*Harposporium anguillulae*), helicoid (*H. helicoides*), or with irregular processes emerging from odd-shaped conidia (*H. diceraeum*). After ingestion, the position within the gut that conidia finally become lodged depends on the shape of the conidium, which is species-specific. For example, conidia can become lodged in the muscle tissue of the oesophagus (*H. anguillulae*), travel to the lower gut (*H. arcuatum, H. helicoides, H. oxycoracum*), or become anchored in the buccal cavity (*H. bysmatosporum, H. diceraeum*). The conidia appear unaffected by being ingested by a nematode and are resistant to the digestive enzymes produced by the animal. The crescent-shaped conidia of the most widely distributed and frequently isolated species, *H. anguillulae,* has one sharply pointed end. This penetrates between the muscle fibres in the oesophagus and once firmly embedded cannot be dislodged. Ultrastructure studies have shown that the sharply pointed distal process of the conidium always has a number of annular striations, which appear to help in lodging the conidium firmly in the oesophageal lumen of the host.[19] A germ tube develops from the centre of the convex side of the conidia, penetrates the muscles of the oesophagus, and invades the host's body. Assimilative hyphae are formed, and as with species forming adhesive conidia, the body contents are digested and conidiophores produced. Those species producing conidia which travel to the lower gut, such as *H. oxycoracum,* can easily pass through the oesophagus even though they are longer than *H. anguillulae.* The conidia accumulate in the lower gut, and each conidium germinates producing a germ tube which penetrates through the microvillous layer of the gut into the body cavity.[18] In contrast to endoparasites in other genera, the conidia of *Harposporium*

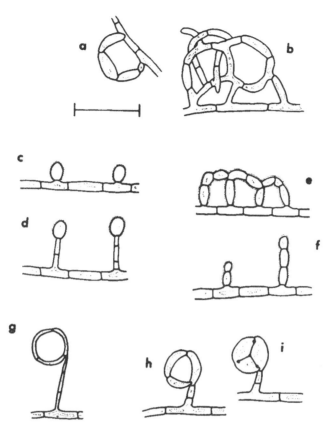

FIGURE 1.   Trapping organs of predatory nematophagous fungi: adhesive nets forming simple (a) and complex (b) three-dimensional traps; sessile (c) and stalked (d) adhesive knobs; and adhesive branches (f) often forming simple two-dimensional adhesive networks (e); nonconstricting rings (g) and constricting rings open (h) and closed (i).

species are supported on spherical or subspherical shaped phialides. Chlamydospores are readily produced by a number of species which allows the fungus to withstand periods of adverse conditions within the soil.

Although the ingested conidia of the genus *Harposporium* do not attract nematodes, the mycelium within the infected nematode is strongly attracting. Jansson[14] has suggested that nematodes are attracted to the vicinity of the infected nematodes and ingest the conidia while they feed. The stylet-bearing plant-parasitic nematodes are unable to ingest the conidia. Adhesive conidia appear to attract nematodes regardless of feeding adaptation. Only one species of the genus *Harposporium* is reported as not infecting nematodes by ingestion. *Harposporium subliforme* has awl-shaped conidia with adhesive apical processes, which allows it to adhere to the cuticle of its prey.[20,21]

B. Predators

In the natural environment, and under the low nutrient conditions of water agar plates used to isolate the group, predatory fungi form extensive but generally quite sparse mycelium. When nematodes are present traps are produced at intervals along the length of hyphae. There are six different types of traps found: undifferentiated or unmodified adhesive hyphae, hyphal branches anastomosed to form three-dimensional adhesive nets, adhesive branches which sometimes link to form simple and usually two-dimensional adhesive nets, adhesive knobs, constricting rings, and nonconstricting rings (Figure 1). The Zygomycetes are able to capture nematodes along the entire length of their undifferentiated mycelium by producing adhesive. However, the ability to produce adhesive in the nematophagous Hyphomycetes is restricted to specialized trapping organs. In temperate soils, adhesive knobs, constricting rings, and three-dimensional adhesive nets, are the most frequently observed trapping mechanisms isolated. They represented 31.3, 28.6, and 22.3%, respectively, of the predatory fungi found in Irish soils.[9] Adhesive branches (9.8%) and unmodified adhesive hyphae (7.1%) are locally abundant while nonconstricting rings are comparatively rare (<0.9%).

The life cycles of predatory fungi are essentially the same, with basic similarities in the trapping and initial penetration of prey, even though the traps are significantly different morphologically.[22] A nematode becomes attached to or is caught in a trap, and the cuticle of the prey is rapidly penetrated by a narrow infection hyphae, which swells inside the animal's body to form a spherical bulb.[23] In some species, such as the genus *Nematoctonus*, an infection bulb is not formed, but in others it can develop until it is the same diameter as the width of the nematode. Assimilative hyphae develop from the bulb, and quickly absorb the contents of the nematode. Nematodes are immobilized fairly rapidly, with the exception of those caught by nonconstricting ring traps, with all movement ceasing within 1 to 2 hr of capture. It has been suggested that nematophagous fungi produce a substance which first immobilizes and then kills the host, a nematoxin, although this has never been firmly established. Various workers have tried to identify the nematoxin and suggestions have included ammonia or a polysaccharide.[24-27] Whatever the material is, presumably it is produced by the trapping organs and as passing nematodes are unaffected the action must be very localized, so the toxin is either in the adhesive or released into the nematode via the infection hyphae. Either way the trap itself must be sufficiently strong to hold onto the nematode for some considerable time without the aid of an immobilizing toxin, so the advantage of producing such a substance to the fungus is unclear. It is interesting to note here that nematodes infected with endoparasites can have quite extensive hyphal growths developed within their bodies and still be quite active, so it would appear that nematoxins are not produced to the same extent in parasites, if at all, compared with predators. However, when the adhesive conidia of *Nematoctonus concurrens* becomes attached to the cuticle of a nematode, immobilization takes place within 24 hr.[28] It was recorded that there was often extensive hyphal growth over the surface of the nematode before penetration occurred. Once all the nutrients have been absorbed the protoplasm in the assimilative hyphae is translocated to the main mycelium. The nematode corpse, filled with empty hyphae, is rapidly degraded by other saprophytic microorganisms. The absorbed nutrients can be used for a number of functions, such as further hyphal growth, production of conidiophores and conidia, or the formation of chlamydospores. Conidiophores are often formed at the sites of capture, with the number of conidia produced varying according to species. For example, the conidia of *Dactylella doedycoides* and *Dactylella bembicodes* are borne singly at the apex of unbranched conidiophores, while the conidia of *Arthrobotrys oligospora* are formed in a succession of clusters or whorls along the length of the conidiophores with up to 40 to 50 conidia per conidiophore being common. The exact nature of the dispersal of the conidia is not known. They are generally quite large, the conidia of *D. bembicodes* for example are typical, with a range 34.0 to 48.0 × 16.0 to 23.0 μm. As they are quite susceptible to desiccation, they are probably not dispersed aerially, but more likely inadvertently dispersed by soil macroinvertebrates. Conidia of this size contain considerable food reserves, and so under suitable conditions the conidia germinate, producing hyphae and traps using internal reserves only.[5]

1. Unmodified Adhesive Hyphae

This trapping strategy is restricted to the Zygomycetes, except for a single Hyphomycete *Arthrobotrys botryospora*, which was reported by Barron[29] as also capturing nematodes in this way, although adhesive nets was the primary trapping device used. Gray[30] has noticed a similar ability in trapping nematodes in the net-forming species *Dactylaria psychrophila*, while Jansson and Nordbring-Hertz[31] described a similar behavior with *A. superba*. This suggested that the use of unmodified adhesive hyphae for catching nematodes may be more widespread than previously thought. The Zygomycetes are aseptate fungi and so are incapble of forming sophisticated trapping organs as seen in the septate fungi. The hyphae is used as the trapping organ with prey captured by adhesion, so apparently all parts of the mycelium are capable of capturing prey. The adhesive is extremely strong and can capture even large nematodes at a single point of contact, unlike net-forming species where more than one point of contact on the adhesive network is usually required to hold large nematodes. The nematode becomes immobilized 1 or 2 hr after capture and is then penetrated by a fine lateral branch. This gives rise to assimilative hyphae which rapidly fill the nematode, absorbing the body contents within 24 hr. The adhesive often has a golden yellow color and so is easily seen with normal light microscopy. It is either secreted over the entire surface of the hyphae, or is only produced at a single point on the hyphae in response to contact with a nematode. In the latter, nematode contact produces a rapid movement of cytoplasmic particles along the hypha, to the point of contact, where a drop of adhesive material is exuded within a few seconds. Of the ten or so species so far isolated, most belong to two genera; those species that produce conidia generally belong to the genus *Stylopage*, while species not producing conidia, only chlamydospores, belong to the genus *Cystopage*. Surprisingly, little is known of this group of predatory fungi, although recent notes and observations on specific species have been made.[32-35] Most surveys have shown this type of tapping mechanism to be infrequent in soil. However, Bucaro[36] examined 38 samples of soil from El Salvador and isolated 18 species of nematophagous fungi, two of which had unmodified adhesive hyphae, *Stylopage grandis* and *S. hadra*. The latter species was the most frequently isolated of all the species found, being present in 21 of the samples and representing 37% of the total records of fungi.

2. Adhesive Branches

The adhesive branch is the next simplest morphological trapping structure from the unmodified adhesive hyphae. Erect branches made up of 1 to 3 cells are produced, which normally anastomose to form simple adhesive hoops or two-dimensional networks which are similar in appearance to a line of crochet (Figure 1e). The entire branch is covered with a thin film of adhesive so that nematodes can be captured if they come into contact with any part of the branch. Adhesive branches are normally closely spaced so that if a nematode becomes attached at one point it will quickly come into contact with other adhesive hyphae as it struggles to escape, ensuring the prey does not break free. In temperate soils *Dactylella cionopaga* is the most frequently isolated species which forms adhesive branches, and this species has been extensively studied.[37]

3. Adhesive Nets

The most common trapping method developed by nematophagous Hyphomycetes is the complex three-dimensional adhesive network. With 35 species known it is frequently encountered in nearly all types of soils. Perhaps the best known species with this type of trapping device is *Arthrobotrys oligospora*, which has a world-wide distribution. Adhesive nets are an evolutionary development from adhesive branches, and are formed by an erect lateral branch growing from the vegetative hyphae, and curving so that it is able to fuse with the original parent hypha. Further lateral hyphae are produced from the parent hypha or from the loop, to form further loops, until a complex mass of anastomosed loops are formed, which develops away from the parent hyphae in all possible directions (Figure 1b). The entire network is coated with a thin layer of adhesive on which nematodes are captured. However, compared with other types of trapping mechanisms they are less aggressive predators, with casual contact with nematodes not always resulting in capture. The design of the nets ensures that as the captured nematode struggles, it becomes entangled in the net and stuck to other parts of the network. As can be seen in aquatic environments where networks have been found, the hyphae comprising the network is more robust than the vegetative hyphae, and often just the nets alone are observed.[38] The traps contain numerous electron-dense vesicles which are not present in the vegetative hyphae, and large osmiophilic inclusions are more common in traps. It appears that the adhesive and the digestive enzymes used for capture and penetration of the nematode, respectively, originate from either or both of these organelles.[39,40] Once captured, a penetration hypha enters the nematode within 1 hr and then swells to form an infection bulb.[23] From this structure assimilative hyphae develop.

4. Adhesive Knobs

Adhesive knobs are morphologically distinct cells, covered with a thin layer of adhesive (Figure 1d). They are either sessile (*Dactylella phymatopaga*) or borne at the apex of a short nonadhesive stalk (*Dactylella ellipsospora*). The adhesive knobs formed by *Dactylella parvicollis* continue to grow to form a short chain of adhesive cells, which can link to form an adhesive ring. As well as adhesive knobs, some species also have nonconstricting rings as an additional trapping mechanism, the most frequently isolated species being *Dactylaria candida*. The adhesive knobs are globose in shape and so when a nematode is captured the area of contact between the nematode and the knob may be very small, resulting in the prey being able to struggle free. The fungus overcomes this in three ways. When the nematode is captured, a flattened mass of adhesive is produced at the point of attachment, forming a thick pad. This increases the area of attachment many fold, thus ensuring the nematode is firmly held.[41,42] Adhesive knobs contain numerous electron-dense spherical bodies close to that part of the cell wall likely to come into contact with a nematode.[43,44] These bodies are similar to those found in other adhesive traps, and are not present in vegetative hyphae. Penetration by the fungus is a mixture of enzymatic activity. For example, collagenase is produced which may aid penetration of the cuticle[45] and physical activity. The extra strength of attachment, provided by the thick, adhesive pad, is required if the penetrating hyphae is to force its way through the cuticle of the nematode. Similar pads are produced by adhesive branches and nets.[39,46] The wall structure of the trap allows the passage of extra adhesive lateral to the point of contact with the nematode. Nordbring-Hertz and Stalhammar-Carlemalm[39] found that in the net-forming species *Arthrobotrys oligospora* this thick deposit was formed within 15 min of capture. Adhesive knobs are normally quite closely spaced along hyphae so when a nematode is caught it is quite normal for it to become attached to several other knobs as it struggles to free itself, making escape impossible. The adhesive knob is separated from the stalk or hyphae by a septum, and Barron[47] observed that the knobs formed by several species were easily detached from their stalk when the nematode struggled, but remained firmly attached to their prey, finally penetrating the nematode as normal. This phenomenon appears common in a number of species, especially those which also form nonconstricting rings.

Adhesive knobs can be formed directly by the conidium without any vegetative mycelium being produced. This confers a significant ecological advantage on the fungus which is able to avoid any fungistatic effects of the soil which may inhibit normal vegetative growth and subsequent trap formation.

There are nearly 20 species which form adhesive knobs, with the trapping mechanism evolving independently in both the Deuteromycetes and the Basidiomycetes. The Basidiomycetes of the genus *Nematoctonus* have very characteristic shaped adhesive knobs (Figure 2). They are comprised of an hour-glass shaped secretory cell, strongly attached to the hypha, either directly or via a short branch, and enclosed in a spherical ball of adhesive. The

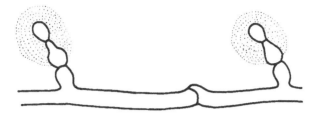

FIGURE 2. Hyphae of *Nematoctonus* showing clamp connection and hour-glass shaped adhesive knobs.

adhesive produced by these species is extremely strong and once a nematode comes into contact it is unable to free itself. Conidia are formed on short denticles off the vegetative hyphae, and readily become detached. Adhesive knobs are formed at the tip of the dispersed conidia to capture prey and complete their life cycle.

5. Nonconstricting Rings

This is the most infrequently encountered trapping mechanism, with only four species of nematophagous fungi known to capture nematodes in this way. Nonconstricting rings are produced by erect lateral branches arising from the vegetative hyphae. The branch is very slender at first, but then thickens and curves to form a three or four-celled ring, which fuses with the support stalk just above the point where it increases in width (Figure 1g). Nematodes are caught by passively entering the ring which becomes wedged around its body. The more it struggles to free itself the more firmly wedged it becomes, often causing a visable constriction around the animal. While the nematode can be held by the ring at the site of capture, very often the nematode breaks free from the mycelium, but the ring remains firmly in place to subsequently invade its prey. The fine supporting stalk can often be observed to have degraded just below the point of contact with the ring, which suggests that the fungus wants the captured nematode to break away with the ring firmly wedged around its body. Like those species producing adhesive knobs which become detached from the main hyphae, this appears to be a deliberate strategy to achieve maximum dispersal within the soil. The most frequently isolated species of this group is *Dactylaria candida*.

Working on the ultrastructural features of the nonconstricting rings formed by *Dactylella leptospora*, Saikawa[48] has shown that such traps may not be passive as previously thought. Electron-dense inclusions positioned towards the luminal side of the ring cytoplasm are clearly visible, the inclusions being similar to those seen in adhesive traps. Also, an adhesive layer, similar to those associated with adhesive nets and adhesive knob traps,[42,49] was observed on the inside of the ring.

6. Constricting Rings

This is perhaps the most aggressive and successful trapping device found in the Hyphomycetes with 12 species known to produce constricting rings. The size of the ring varies between and within species, but generally fall within a range of 20 to 40 μm internal diameter (ID), while several species have been reported as producing giant or macro-rings (>90 μm ID) on nutrient agar, although the reason for this is unclear. Megatraps have been induced in strains of *Dactylella brochopaga* by Insell and Zachariah[50,51] who have found that they are able to function quite normally. The fungus achieves this by reducing the length to width ratio of the cells making up the constricting ring, and Insell and Zachariah[50] have been able to quantify this relationship. However, whether this is an adaptation for capturing larger and more slowly moving prey is not known. Like nonconstricting rings, constricting rings are a mechanical trap and are nonadhesive. They are made up of three arcuate cells which form the ring that is attached to the hypha by a short stalk (Figure 1h, i). Nematodes entering the rings stimulate the inner surface of the cells which, after a short lag period of 2 to 3 sec, rupture and close around the prey holding it fast within the ring. Present knowledge is restricted to the basic facts that the ring closes extremely rapidly (<0.1 sec), that closure is stimulated by tactile and heat stimulation, and that rings can function even when detached from the parent hyphae. However, the exact mechanism of trap closure is still unknown, even though it has been the subject of intense research and speculation.[52-58] Barron[5] has recently proposed a simple, and in many ways a more plausible, theory of ring closure than previously devised. He compares the structure of the ring to the other fungal cells which rupture at a critical moment in time, such as the explosive dispersion of spores seen in some fungi. A line of weakness exists along the inner wall of the cells making up the constricting ring. It is in such a critical state that any physical disturbance will cause the line of weakness to break and the cell to rupture, increasing the cell wall area by 50%. The cell wall is freely permeable so there is a rapid uptake of water causing the elastic inner wall to balloon inwards, constricting the nematode as the cell swells to three times its former volume. The mechanism of constricting ring traps is fully examined by Insell and Zachariah,[58] and the closure of traps is illustrated using unique time-lapse photographs.

The formation of constricting rings is similar to nonconstricting rings although somewhat more complex. A lateral branch is formed from the vegetative hyphae and grows in a curve. The loop of the curving branch is attracted to a bud formed near the base of stalk onto which it anastomoses. At this time the cells of the ring become distinct and the advancing tip of the curved branch also anastomoses with the base of the first cell of the newly formed ring, uniting the first and last cell of the ring. The ring quickly matures and is then ready to capture a nematode. The formation of the ring has been described in detail by Drechsler[59] and illustrated by a sequence of remarkable photographs of *Arthrobotrys dactyloides* by Higgins and Pramer[60] and more recently of *Dactylaria brochopaga* by Barron.[5]

The ultrastructure of the ring cells, and other trapping organs, differs considerably from the vegetative cells of nematophagous fungi.[55,61] In open constricting ring cells there are membrane-bound inclusions, labyrinthine matrix, and electron-lucent regions between the protoplasts and cell wall, all localized on the luminal side of the ring cells. After closure, the labyrinthine matrix stretches becoming smooth, and the electron lucent layer and the other features disappear. This action is irreversible and it is impossible for rings, once inflated, to deflate and be reused.

III. IDENTIFICATION

A. Taxonomy

It is not possible to identify nematophagous fungi by the conidia alone, the trapping mechanisms also an important diagnostic factor. Most of the isolation techniques used for the group rely on the fungus making itself known by capturing or infecting nematodes. In pure culture, many of these fungi do not produce trapping organs, and the dimensions of the conidia and conidiophore are often different to those examined on baited plates. Because of this, the characters used to identify these fungi are those found on nematode-infected cultures. An early key by Dollfus[62] was extensively revised by Cooke and Godfrey,[63] and while this contains details of all the commonly found species, only 97 species are included, so another 50 to 60 species are not covered. Since 1964 many new species have been described, and although many are either scarce or of restricted distribution, some species such as *Arthrobotrys pauca* may be far more abundant than previously thought. The problem with any key is that unless constant reference is made to the original description of fungi, the user will generally opt for the ''best fit'' approach to identification. While Cooke and Godfrey's key is undoubtedly the best available at the present time, and is still widely used,

Step

1   Endoparasites with assimilative hyphae................2. (conidia-forming species)
     Endoparasites with vegetative or hyphal bodies.......26. (mainly species producing zoospores)
     Predators..33.

6   Hyphae with clamp connections - Present.............7. (<u>Nematoctonus</u> - adhesive cells formed)
                                       Absent.............12.

12   Conidia borne on - Sterigmata......................13. (<u>Meria</u>)
                          Phialides......................14.

14   Conidia - arcuate, slightly curved, straight generally on subspherical phialides.....15. (<u>Harposporium</u> - ingested
                                                                                               conidia).

        oblong, ellipsoidal, spherical generally on flask-shaped phialides.........21.   (<u>Verticillium, Acrostalagmus,</u>
                                                                                         <u>Cephalosporium</u> - adhesive conidia).

33   Unmodified hyphae....................................34. - Aseptate.....35.
                                                                 Septate......38.

     Hyphal traps formed..................................39. - Aseptate.....40.
                                                                 Septate......41.

41   Adhesive branches......42.
     Adhesive knobs.........43. - Conidiophore usually branched near apex, knobs always branched ....44.
                                   Conidiophore usually simple, knob sessile or branched..................45.

     Non-constricting rings..50.

     Constricting rings......54. - Conidia borne in terminal cluster........55.
                                    Conidia borne singly.....................58.

     Adhesive nets...........63. - 1 septa.................................64.
                                    >1 septa................................72.

64   Conidia - Not truly capitate...65.
                 Truly capitate..66.
                 Loosely capitate..71.

72   Single conidium.........75.
     Cluster of conidia......77.

FIGURE 3.   Major steps in the identification key of Cooke and Godfrey.[63]

a new key is required which will not only update the taxonomy and include the numerous species not included in the present key, but will also review certain groups.

For those who are not able to expand the key of Cooke and Godfrey and develop their own key, the former is still extremely useful and is recommended for general identification. The main steps of the key are summarized in Figure 3; however, there are a number of points in the key where the nonspecialist may have problems:

Table 1
SUMMARY OF THE MAJOR MORPHOLOGICAL FEATURES OF
ACROSTALAGMUS OBOVATUS AND *CEPHALOSPORIUM BALANOIDES*[64]

	Acrostalagmus obovatus	*Cephalosporium balanoides*
Assimilative Mycelium	Hyaline and filamentous Branched and septate 2.00—3.00 μm wide	Hyaline and filamentous Branched and septate 1.5—3.0 μm wide
Conidiophore	Creeping, rarely ascending Simple or sparingly branched 50—500 μm in length 1.7—3.0 μm wide Cells 15—20 μm long	Creeping, rarely ascending Simple or sparingly branched 50—500 μm in length 1.2—2.5 μm wide Cells 10—30 μm long
Phialides	Flask shaped Some cells bearing 1 to 6 phialides 6—10 × 2.5—3.5 μm Distal sterigma 0.5 μm wide Phialides borne singly or in pairs, numbers from 3 to 6 formed in whorls or irregular groups	Flask shaped 1 or 2 phialides per cell at the distal end 7—20 × 2.0—3.5 μm Distal sterigma 0.5—0.8 μm wide Phialides irregularly formed
Conidia	Aseptate Ellipsoid or obconical 3.0 × 2.0 μm Conidia in cluster at apex of sterigma Up to 20 conidia	Aseptate "Acorn-shaped" 2.4—2.8 × 2.3 μm Conidia in cluster at apex of sterigma Up to 15 conidia

1. The separation of *Acrostalagmus obovatus* and *Cephalosporium balanoides* is difficult as they are so similar at magnifications up to 400 and even higher (stage 24 in Cooke and Godfrey's key). This is particularly irritating as the two species are amongst the commonest endoparasites of nematodes. Notes on the separation of these two species are given by Gray[64] which is summarized in Table 1.

2. The separation of species producing zoospores is extremely difficult as the life cycle of the fungi is completed so quickly and the spores rapidly dispersed (stage 29 of the key). However, *Protascus subuliformis* has been transferred to the genus *Myzocytium* by Barron[65] who has devised a key to the species of *Myzocytium*. In order to identify these species they really need to be mounted on slides, stained, and examined at magnifications of at least a 1000.

3. For students, stage 64 in the key is often the one that misleads them the most. Here the net-forming species producing conidia with a single septum, which includes the majority of species, are separated according to whether the conidia are truly capitate, loosely capitate, or not capitate. The bulk of known predators fall into one of these three categories and so it is important to make the correct selection at this point in the key before preceding. This is clarified in Figure 4.

4. There are many variations in the size range and shape given for conidia[66] and really there is no substitute to referring back to the original description of the fungus. Also, care must be taken when it states in the key that a predatory species does not produce chlamydospores. For example, in the key, the absence of chlamydospores in *Trichothecium cystosporium* (stage 65) and *Arthrobotrys superba* (stage 68) are quite important diagnostic features, and so if chlamydospores are present than these common species tend to be immediately ruled out. The professional mycologist will smile at this point and say that of course chlamydospores are possible in any Hyphomycete,

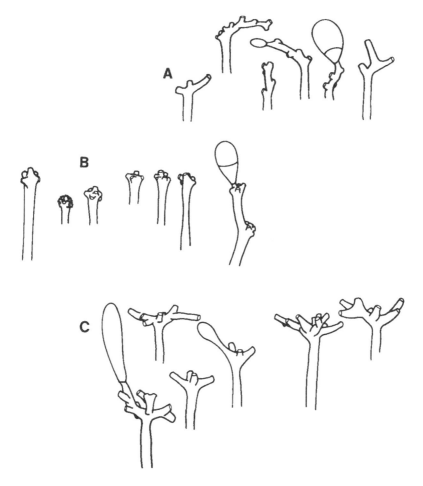

FIGURE 4. Difference between the degree of capitation used to differentiate between net forming nematophagous fungi at stage 64 of the key devised by Cooke and Godfrey.[63] (A) Stage 64. Conidia not truly capitate[65] — formed in basipetal succession; forming a panicle; elongation between conidia. Example: *Trichothecium cystosporium*. (B) Stage 64. Conidia capitate[66] — formed in close heads; type *Arthrobotrys*. Examples: *A. superba* (net-former), *A. dactyloides* (constricting ring former). (C) Stage 64. Conidia loosely capitate[11] — not formed in panicle; "candelabrum"-like branching apex; capitate but nodes at apex are slightly elongated. Examples: only one species known from Europe (*A. musiformis*), *A. javanica* is known from Java. Proposed as separate genera — *Candelabrella* now in *Arthrobotrys*.

and so one should not put too much reliance on the fact the key or the original descriptions of the fungi state otherwise. However, I personally find it misleading.

Much confusion exists regarding the taxonomy of many of the nematophagous fungi. *Meria coniospora*, a very common endoparasite has recently been transferred into a new genus of Hyphomycetes called *Drechmeria*.[67] The nematode-parasitic species of *Verticillium*, *Acrostalagmus*, *Spicaria*, *Cephalosporium*, and *Cephalosporiopsis* are extremely similar, and it would seem sensible to include them in a single genus *Verticillium*.[68] *Cephalosporium balanoides* and *Acrostalagmus obovatus* have both recently been described and transferred to *Verticillium*,[15] and *Verticillium sphaerosporum* appears to be synonymous with *A. obovatus*. The whole genus *Verticillium* is at present under review which should clarify the present position enormously.[69] Due to the variability in conidial dimensions from different locations, or when grown under different nutrient conditions, there seems little justification for the separation of species due to minor differences in conidia size or shape.[70,71] It may well be that the species of the genera *Harposporium* and *Myzocytium* for example, should be regarded as phenomorphs, and included under single specific epithets.

In comparison with the endoparasites, the taxonomy of the predators has been extensively investigated and revised. Schenck et al.[72] transferred all the nematode-trapping species of *Dactylaria* and *Trichothecium* to the genus *Arthrobotrys*. They also demonstrated that *Didymozoophaga*, *Genicularia*, *Candelabrella*, *Dactylariopsis*, and *Duddingtonia* were all synonymous with *Arthrobotrys*. Most of the other species, mainly of the genus *Dactylella*, have been transferred to *Monacrosporium*.[73] Schenck *et al.*[72] did not tansfer the species in *Monacrosporium* to the genus *Arthrobotrys*. However, Tribe[74] argues that these species are no more distant from the type species of *Arthrobotrys* than are some species subsumed under *Arthrobotrys* by Schenck et al.[72] While there is still considerable confusion regarding the generic identity of so many species of nematophagous fungi, the species names used in this paper are those used by Cooke and Godfrey,[63] or subsequently given in original descriptions of new species, so as not to confuse the reader. However, it is strongly recommended that those regularly using this key should transfer the typescript onto a microcomputer, update the generic names of the fungi using Cooke and Dickinson[73] and Schenck et al.,[72] and then gradually add extra information of individual species as collected. The only way to compile an accurate working key is to compile your own file of photographs and drawings, and of course by sending material to the major culture collections or to other workers for verification.

Like the endoparasites, predators show a large degree of variation within species, which often makes it difficult to match them perfectly to the often very precise original descriptions. Isolates of predatory nematophagous fungi show a definite gradation of morphological characteristics, as well as physiological and biochemical characteristics, as one would expect in microorganisms exposed to varying environmental conditions. So there is a dilemma for the taxonomist trying to separate out species essentially using morphological characters only, especially when these very characters are so changeable with environmental factors such as nutrient status. Yet it would be both impossible and unwise to try to separate these isolates and assign species status to them. Those who regularly try to identify nematophagous fungi to species level will most likely agree with Mankau[1] who suggests that a genus such as *Arthrobotrys* may in fact be as varied and complex as *Penicillum* and its relatives.

IV. ISOLATION

Most nematode pathologists will have come across nematophagous fungi while directly examining nematodes under the microscope. Endoparasites are the most likely fungi to be seen by direct observation, with nematodes filled with assimilative hyphae and perhaps short conidiophores growing out from the host with conidia attached. Often large hyphal bodies are observed within nematodes of the endoparasitic Chytridiomycetes and Oomycetes, or detached trapping organs of the predators such as adhesive knobs or nonconstricting rings adhered to the animal's body. If these infected nematodes can be separated then they can be transferred directly to agar plates to be cultured and the infecting fungus eventually identified.

In the soil nematophagous fungi are present as infected nematodes, vegetative or predatory hyphae, and conidia. So the isolation techniques used are aimed at the recovery of some or all of these viable fungal components. The isolation techniques used for the isolation of these fungi from source material, such as soil or associated vegetation, rely on three principles:

1.   The use of samples of soil plated directly onto a low nutrient agar which is baited with nematodes. The low nutrient status of the agar prevents the fungi growing saprophytically and forces them into a predatory mode using the nematodes as the source

of nutrients. Unwanted fungal and bacterial contamination on the plates is prevented by using low nutrient agar, which reduces the competition from saprophytes, allowing the nematophagous fungi to be more readily observed. This method is used to primarily recover predators although endoparasites are also isolated.

2.   Nematode extraction is the main method used for isolating endoparasites. A variety of techniques can be used to remove the indigenous nematode population, a proportion of which will be infected with any endoparasites present in the soil. The process allows quite large samples of soil to be processed, with the extracted nematodes plated onto low-nutrient agar to complete their life cycle. Occasionally predators may be recovered from nematodes which have adhesive trapping organs attached to their cuticle, but this is almost exclusively restricted to adhesive knob species.

3.   The final method uses a separation process to remove the viable fungal component from the soil. This is achieved by making the soil into a slurry by adding water, then separating the fungal fraction and the nematodes from the soil particles by settlement or filtration. Both predators and endoparasites can be recovered, and as such processes allow much greater volumes of soil to be processed than the direct soil plating techniques, a better recovery of species is possible. However, these methods are far more complex and time consuming than the others,[75] and as no comparative study has been made on their relative efficiencies, the majority of workers in the field have continued to use techniques using the first two principles.

It is not known whether nematophagous fungi are evenly distributed throughout the soil or whether they are restricted to particular zones, as are nematodes. However, high levels of activity of the fungi are found in the rhizosphere area of plants,[76,77] and for general isolation work it is convenient to take subsamples from this distinct area whenever possible. Where no distinct zones exist then the material is gently mixed with a pestal and mortar, and subsamples taken randomly from the homogeneous samples. The most effective isolation methods in terms of recovery of species and time are the soil sprinkling and the Baermann funnel techniques.[78]

In the soil sprinkling technique, approximately 0.5 g of fresh sample is sprinkled over the surface of freshly prepared 2% water agar plate. Although the plates can then be left for several days before being subsequently inoculated with a concentrated nematode suspension, it is best to inoculate the plates with nematodes at the same time as the soil is added. However, the addition of the nematodes a few days prior to adding the soil does have the effect of reducing any subsequent contamination of the plates by bacteria and myxomycetes. Wyborn et al.[79] found that maximum recovery of species was obtained by using an inoculum of approximately 5000 nematodes per plate. Nematode inocula in excess of this results in the surface of the agar being churned up making observations of the fungi difficult. Inocula of less than the recommended number reduces the level of stimulation so that not all species are recovered.

Nematodes for baiting plates can be easily cultured or collected using a soil extraction technique. The nematode most widely used for studies with nematophagous fungi is *Panagrellus redivivus*, which has been shown to be an acceptable target organism for a wide range of such fungi.[80] Cultures are maintained by adding an inoculum of *P. redivivus* to sterilized moist oatmeal in a 150 mm diameter glass Petri dish, which is then incubated at room temperature. After 7 to 10 days a dense culture of nematodes has developed, many of which will collect on the underside of the Petri dish, which can subsequently be collected by simply washing them off. Heintz[81] removed nematodes from the surface of oatmeal cultures using sterile swabs and transferring them to sterile water. Once separated, the nematodes should be washed several times in sterile water before use. Fresh oatmeal cultures must be set up regularly as once the nematode culture becomes old the inocula fail to recolonize fresh medium. Barron has extensively used *Rhabditis terricola*, which was originally isolated from rotting potatoes.[82] The nematode is maintained on Nigon's medium,[83] although the population density never reaches the levels seen with *P. redivivus*. Barron found that an improved method of culturing this species was to centrally inoculate a water agar plate with 1 mℓ of the nematode suspension and then sprinkle the area with 0.5 g of autoclaved Lipton's® green pea dried soup. Plates were inoculated at either 20 or 30°C to obtain large populations of nematodes within a few days. An alternative to maintaining a culture of nematodes, but much more time consuming, is to use the indigenous population of nematodes found in the soil. This is possible using one of the extraction techniques such as the Baermann funnel method, which is described below.[84] The problem with using soil nematodes is that some of them will be infected with various endoparasites. If this method is used for a semiquantitative study the nematodes should, if possible, be extracted from the same soil that is being plated out. The nematode density in soil can vary widely, for example from 0 to 2086 nematodes per gram in Irish soil,[85] so it is important to ensure that sufficient nematodes are extracted to guarantee an inoculation equivalent of at least 4000 nematodes per plate.

The method most widely used to extract the soil nematode population in order to isolate endoparasites is the Baermann funnel technique. The nematodes are extracted by wrapping approximately 30 g of fresh sample, comprised solely of soil, in a double layer of soft tissue. This is then placed in a standard Baermann funnel apparatus,[86] and the water level readjusted until it comes into contact with the sample. The nematodes wriggle through the tissue, sink down into the base of the funnel, and are collected in a small tube. After 24 hr the collecting tube is removed and the nematode suspension concentrated to 1 cm³, which is then poured onto week-old water agar plates. Endoparasites, if present, then develop within their nematode hosts and can be subsequently identified. Care should be taken to ensure that the apparatus, especially the rubber tubing connecting the funnel and the collecting tube, is thoroughly cleaned to prevent contamination of future samples. Some species of endoparasites may release flagellate zoospores in the collecting tube, so a high incidence of these species on the agar plate will not necessarily reflect an equal abundance in the original sample. The methods used in the isolation of nematophagous fungi has been reviewed by Barron[4] and Gray.[78]

The presence and subsequent identification of nematophagous fungi are normally established by direct observation, locating the fungi by looking for dead or dying nematodes and scanning for the characteristic conidiophores of the predators just above the surface of the agar. Plates should be scanned at ×100 magnification for the presence of both endoparasites and predators. Lower magnification, such as ×40 or ×60 magnification, should only be used by experienced workers, as endoparasites are readily overlooked. Subsequent identification is done directly at ×400 magnification using an objective with a long working distance. The entire surface of each plate should be scanned in this way using a template fixed to a movable stage. Shepherd,[87] used a series of four diagonal scanning lines which all ran through the center of the plate at an angle of 60° to each other, finally scanning around the entire outer edge of the agar. She felt that this method ensured the majority of nematophagous fungi would be observed. However, it appears from comparing this method with the more extensive scanning technique, that although all the predators producing extensive mycelia are isolated, many of the less abundant endoparasites and less virulent predators are overlooked. The time taken to completely scan the surface of an infected plate under the microscope depends on the amount of debris, the total nematode number, and the condition of the agar surface. With experience, scanning a plate containing no fungi can take between 20 to 30 min/plate to upwards of 3 hr when species are present and require identification. So, it is not advisable to set up too many plates at the same time, especially as the emergence time for nematophagous fungi vary considerably. In order to recover the majority of species from a particular sample, the soil plates should be scanned frequently. The life cycle of many endoparasites take only a few days to complete and Barron[82] advises that plates should be examined daily to ensure that all the possible endoparasitic species are observed. While the plates containing the Baermann funnel extracts should be examined daily for 10 days and then at weekly intervals for 8 weeks, the soil plates should be examined as frequently as possible, and at least at weekly intervals, for 10 weeks. In this way even the most slowly developing species will be recovered, but less frequent observations will result in many fewer species being recorded.

Direct observation at × 400 magnification using an objective lens with long working distance has proved a most successful method of identifying nematophagous fungi. It significantly reduces the number of mounted preparations that need to be made, thus saving considerable time. When observed fungi are immature, the agar is marked using a scalpel so that the specimen can be relocated and examined at a later date. Alternatively, where immediate identification is not possible, cultures should be prepared. Shepherd[87] used high-power microscopy to identify fungi directly by adding a drop of water to the agar, lowering a cover-slip over the specimen, and examining it under oil immersion. However, this is rather destructive to the plate and should only be used as a last resort. Wood[71] examined his isolates in temporary water mounts or in lactophenol acid fuschin, while live nematodes were fixed in 5% formalin containing acid fuschin. It is important to retain the hyphae *in situ* for subsequent examination and so permanent preparations can be made by the method originally described by Dixon and Duddington.[88] A thin section of the agar containing the fungus is removed using a scalpel and transferred onto the slide. The agar thickness is reduced by gently warming the slide, and the specimen is stained using cotton blue in lactophenol. Staining fungi on the agar plate is not really necessary for identification and generally kills the specimen, however it does produce excellent photographs. For color photography the stain used by Faust and Pramer[89] is probably the best with the fungal walls stained dark green to blue, the traps taking up the stain most intensely, and the captured nematodes stained bright yellow. They used the monazo dye Janus Green at a concentration of 0.01% (w/v) in 0.2 M sodium acetate-acetic acid buffer at pH 4.6. It is added drop by drop directly to the surface of fungi growing on the agar.

While the predatory Zoopagales have not been obtained in pure culture so far, predatory Hyphomycetes are relatively straightforward to isolate into pure culture. Their conidia are large enough to be collected from the agar plate and their mycelium robust enough to be transferred onto fresh medium by removing a segment of agar. Aschner and Kohn[90] recommended that the tip of a fine mounted needle could be used for conidial transfer. The needle is first flamed and then cooled in agar so that the conidia will readily adhere to the needle. Using a microscope at × 60 or × 100 magnification the needle is used to carefully collect conidia directly from the conidiophores. Alternatively, a single-haired sable paint brush can be used, which allows much more delicate control under the microscope. The collected conidia are streaked onto nutrient agar (corn meal agar) and stored at room temperature. There is little contamination as the conidia are not collected from the surface of the agar. However, if pure cultures are not obtained at this stage, subsequent transfer onto fresh medium is normally required. Then, by transferring a small segment of agar with the fungus onto water agar and adding an inoculum of nematodes, the fungus will produce traps and conidia allowing positive identification.

Endoparasites are more difficult to transfer and grow in culture. The simplest method is to remove the infected nematode with a needle and wash it several times in saline to remove any contamination. This is best done by using a series of cavity slides and transferring the nematode from slide to slide. The infected nematode is then placed on a fresh water agar plate and inoculated with nematodes. The infection cycle normally takes between 2 to 8 days so numerous nematodes rapidly become infected. Giuma and Cooke[91] outlined a centrifugation technique for washing the nematode host to reduce contamination. Aschner and Kohn[90] described how pure cultures can be obtained by transferring nematodes which are infected, but with the fungus not yet sporulating, directly onto nutrient agar (glucose-yeast-extract) supplemented with aureomycin (30 g/ℓ) to reduce bacterial contamination. The parasites, instead of producing conidiophores and conidia, produce vegetative hyphae which continue to grow over the medium. Subsequent transfer by hyphal tipping can then produce a pure culture. While cultures of Hyphomycetes are possible, the nonseptate endoparasites, with the exception of the *Catenaria* species, have not yet been obtained in pure culture. A word of caution however, obtaining endoparasites in pure culture is rather difficult, and those attempting it should not be discouraged by a low success rate. The endoparasite *Meria coniospora* is relatively easy to obtain in pure culture and those who wish to try the technique are recommended to try this species first.

V. PHYSIOLOGICAL CONSIDERATIONS

A. Nutrition

It is surprising that although nematophagous fungi have evolved this specialized habit of utilizing nematodes as a food source, they have retained similar nutritional requirements as other saprophytic fungi when in pure culture. They are able to utilize a wide range of simple carbohydrates as the sole source of carbon and a wide range of nitrogen compounds as the nitrogen source.[92-95] While they all grow well in the absence of prey, differences in the ability of certain predatory Hyphomycetes to utilize various carbon and nitrogen sources have been observed by Satchuthananthavale and Cooke.[95-97] While comparing net-forming and constricting ring fungi they discovered that constricting ring species were much less versatile nutritionally, being unable to utilize carbohydrates more complex than hexoses. Constricting ring fungi also had a reduced ability to utilize inorganic nitrogen but showed a high dependence on organic nitrogen as their major source of nitrogen. In contrast, net-forming species were able to use a wide range of carbon sources and both inorganic and organic nitrogen. The net former *A. oligospora*, which has a high saprophytic ability, was able to utilize all the carbohydrates tested including a range of polysaccharides such as cellulose, starch, and glycogen. *D. bembicodes*, a constricting ring trap former with a poor saprophytic ability,[98] was unable to utilize either cellulose or starch, and was very inefficient at utilizing glycogen, maltose, or sucrose.[95] These results provide a vital clue to the mode of existence of these trap-forming predators in the soil, with the loss of nutritional versatility seen in constricting ring species associated with a higher dependence on the predatory habit. With a limited ability to utilize soil nutrients, constricting ring species are restricted in the soil compared with the net-formers, being dependent on nematodes for the bulk of their essential nitrogen and possibly carbon as well. So it appears that some net-forming species exist as soil saprophytes, even if nematodes are available for capture, and are able to utilize even long-lasting substrates such as cellulose as energy sources; whereas, constricting ring formers live predaceously in the soil, only reverting to a saprophytic existence when ephemeral carbohydrates such as hexoses are available as energy sources. Strains of certain predatory fungi have been isolated which cannot be subsequently subcultured in the absence of nematodes,[53,99] even though other strains of the same species readily grow on nutrient medium. Such fungi appear even more restricted nutritionally than those producing constricting rings, and would apparently have an even more obligately predatory existence in the soil. Little is known of the nutritional requirements of endoparasites although *Nematoctonus haptocladus* and *H. anguillulae* will grow on simple media containing mineral salts and glucose.[68,100]

B. Host/Prey Specificity

Very little is known of the specificity of particular endoparasitic or trapping nematophagous fungi to particular nematodes. Apart from the detailed observations made by Drechsler from 1933 onwards on the specificity of new species to their nematode hosts or prey, few studies have examined specificity. Certain nematophagous fungi have been observed to confine their attack to a single species of nematode, whereas in other cases, a range of nematodes were attacked.[4] Where more than one species was attacked, the fungus sometimes showed a greater virulence against one species than another. Barron suggests that these results are not unexpected, and fall in line with host-parasite and predator-prey relationships in other groups of organisms, where broad and narrow specificity, and degrees of virulence are well documented. There is a lack of critical evidence to support specificity of nematophagous fungi to particular hosts. Most endoparasitic species are known to have a wide host range. For example, *Catenaria* species are capable of attacking a wide range of nematodes as well as mites, rotifers, and tardigrades.[101] *Catenaria vermicola* is able to attack 11 species of plant parasitic nematodes although ring nematodes (Criconematodeae) were mostly resistant to infection.[102] *Catenaria anguillulae* also attacks a wide range of nematodes, being able to infect species from 13 genera, although species from a further 11 genera were not infected.[103] The zoosporic endoparasites appear nonselective in their hosts with *Xiphinema rivesi* and *X. americanum*, both common parasites of apple and peach trees as well as vectors for tomato ringspot virus, parasitized by a range of such fungi.[104] A new endoparasitic Hyphomycete, *Hirsutella heteroderae*, has been recorded as attacking the hop cyst nematode *Heterodera humuli*. This new species displays a certain host specificity, attacking *Heterodera* and other tylenchid and rhabditid species, but failing to utilize dorylaims and plectids as hosts.[105] A similar close relationship exists between the endoparasite *Hirsutella rhossiliensis* and the nematode *Criconemella xenoplax* which is parasitic on peach trees.[106,107] Endoparasites with adhesive conidia are known to attack a wide range of soil, plant- and animal-parasitic nematodes[16,17] and so appear to be generally nonspecific in their host. Wyborn, working on the selectivity of the endoparasitic genus *Harposporium*, found that significantly fewer stylet-bearing nematodes of several genera were infected than rhabditoid nematodes. He suggested that this is because the conidia of the *Harposporium* spp. are ingested and germination takes place within the nematode as opposed to the other endoparasites.[68] The stylet bearing plant-parasitic nematodes cannot ingest conidia and so are generally not attacked by these fungi, although a nematode of the genus *Aphelenchoides* has been recorded infected with *Harposporium oxycoracum*.[108] Barron[82] carried out an extensive study into the endoparisites of the nematode *Rhabditis terricola*. He found that 32 of the 40 endoparasites isolated from a range of soils could parasitize *R. terricola* indicating that these parasites had a relatively wide host range. Barron suggests that each parasite exists in the soil as a complex of many strains which differ in their ability to attack a given host. So where a range of hosts are available a greater recovery of species should be achieved. This was tested by Gray[80] who compared the recovery of 12 endoparasitic and 18 predatory nematophagous fungi using a range of indigenous soil nematodes and a monoculture of the nematode *Panagrellus redivivus*. The results indicated that both endoparasites and predators displayed a small degree of selectivity and that any selection was most likely due to the anatomy of the nematode and the mode of infection or capture by the fungus. This is confirmed by Pandey[109] who demonstrated that ten trap-forming predators could attack the larvae of animal-parasitic trichostrongylid nematodes *Trichostrongylus axei* and *Ostertagia ostertagi*. Much of the recent research on selectivity has been done using *P. redivivus*[110-112] although it is clear that fungi with adhesive and attracting mycelia or conidia are able to infect bacterial-feeding, fungus-feeding, and plant-parasitic nematodes irrespectively.[17] Predatory nematophagous fungi have been shown to be able to capture a wide variety of nematodes, the adhesive organs and constricting rings seemingly unselective in their choice of prey.[4] For example, *A. oligospora* has been reported as catching a wide variety of nematode genera including free-living soil nematodes, plant parasitic species, and animal-parasitic species such as *Haemonchus contortus, Trichostrongylus axei, Ostertagia ostertagi,* and the *Cooperia* spp.[109,113-117] While some nematodes seem unaffected by the nets of certain fungi, Gray[80] did not observe any selectivity by predatory species, with several predators capturing a variety of nematode genera along the same length of hyphae. Gray recorded that fungi with adhesive mycelia and nets, with many points of contact with its prey, were able to capture and consume larger nematodes than fungi with adhesive knobs only, which often relied on a single point of contact to hold prey. However, all the endoparasites were able to capture nematodes irrespective of size.

Rosenzweig et al.[118] tested the ability of seven predators to capture nine different nematodes which included *Panagrellus silusiae, Caenorhabditis briggsae, Meloidogyne incognita, Pratylenchus penetrans, Bursaphelenchus lignicolus, Neoaplectana carpocapsae, Tylenchorhynchus* sp., *Hoplolaimus* sp., and *Xiphinema* sp. They found that the fungi displayed no selectivity at all, with each fungus able to trap and consume all the different nematodes tested. Nordbring-Hertz and Mattiasson[119] found that attachment of nematodes to the adhesive nets formed by *A. oligospora* was due partly to the presence of a lectin on the traps which bound to a carbohydrate on the surface of the nematode. However, the lectin did not show strict specificity for particular carbohydrates, and also readily bound with red blood cells from different blood groups which have different polysaccharides on their surface. Research to date indicates that in the laboratory lectins produced by certain fungi are in fact specific for a carbohydrate on the nematode cuticle, with carbohydrate specificities determined for the trap lectins of the following fungi: *A. oligospora* (N-acetylgalactosamine), *Dactylaria candida* (2-deoxyglucose), *Meria coniospora* (sialic acid), *A. conoides* (glucose/mannose), *Monacrosporium eudermatum* (L-sucrose), and *M. rutgeriensis* (2-deoxyglu-ljcose).[16,17,116,119-122] However, while lectins appear to be highly specific in binding to their carbohydrate in the laboratory, the fungi do not appear to be specific in their choice of prey. Rosenzweig et al.[118] have investigated this problem and concluded that fungal trap lectins demonstrate a "best fit" specificity towards carbohydrates. They also state that the interaction between the lectin and any given nematode could also be mediated by other carbohydrates and/or other binding forces. So the exact nature of these interactions and their role in host/prey specificity remains unknown. This lectin-carbohydrate binding process appears specific for certain groups of animals. The other soil microorganisms, such as bacteria and fungi, with the exception of the yeast *Saccharomyces cerevisiae*, do not bind to the traps and so do not interfere with or prevent nematode capture.[123] The biochemical background to recognition and host-specificity is discussed fully by Jansson and Nordbring-Hertz in Chapter 3.

In biological control studies attention has been focused on plant-parasitic nematodes, yet when associated with the plant these nematodes are most likely free from most fungal antagonists. However, when they are deprived of their proper host or are migrating to a new host in the soil then they are exposed to fungal attack. It is clear that other free-living nematodes are normally more abundant in the soil, and it is with these species that nematophagous fungi will normally interact. This is illustrated by the evolution of endoparasitic species of *Harposporium* and similar genera which infect nematodes by producing ingested conidia which plant-parasitic nematodes cannot ingest.

The close evolution between a fungus and a nematode is demonstrated by a new nematode-trapping fungus *Arthrobotrys ellipsospora*, which has been recently isolated from pine sap collected from below the bark and cambium layer in a *Pinus*, some 1.5 m above the ground.[124] The fungus, which has unusual adhesive branches, feeds on the pine wood nematode *Bursaphelenchus xylophilus* found in the sap and which is a major pathogen of pine trees.

C. Chemotaxis

There has been considerable observational evidence to support the idea that nematodes are attracted to the mycelium of nematophagous fungi. Until recently however, most of the research had involved the use of filtrates of fungal cultures, and not living mycelium.[25,125] Field and Webster[126] tested five nematophagous fungi with different traps and found that bacterial and fungus feeding nematodes were attracted towards the mycelium only when traps had been produced, and so concluded that attraction was dependent on the presence of traps. However, Jansson and Nordbring-Hertz[110] designed methods not only to detect attraction or repulsion of nematodes by living nematophagous fungi, but also to assess the attraction intensity. They found that while most nematophagous fungi attracted nematodes, one net-former, *A. arthrobotryoides*, actually repelled nematodes. The presence of traps is not always necessary to attract nematodes; however, the presence of traps in cultures of *A. oligospora* increased the ability to attract nematodes by a factor of two.[112] Attraction intensity was lowest in fungi with moderate to high saprophytic ability but increased with increasing predacity or parasitism. The conclusion was made that attraction intensity reflected the dependence of the fungi on nematodes for nutrients.[110,127] Jansson and Nordbring-Hertz tested 23 fungi in all, 9 of which were nonnematophagous species, and of these 9 species 5 were also able to attract nematodes. So the nematode-trapping fungi with moderate to high saprophytic ability have a similar attraction ability as some nonnematophagous fungi. This indicates that an ability to attract nematodes above normal saprophytic fungi is only found in spontaneous trap formers species and endoparasites. There is some evidence to suggest that the response of nematodes to being attracted to nematophagous fungi may differ according to its feeding habits. For example, mycophagous nematodes are particularly attracted to all types of nematophagous fungi[127] and are readily captured by a wide variety of fungi.[128,129] Several factors such as carbon dioxide and various organic and inorganic substances have been thought to be involved in nematode attraction.[4] The study by Jansson and Nordbring-Hertz[110] indicated it was a volatile or a small rapidly diffusing compound which was continuously produced in nonspontaneous trap formers (NSTF), while spontaneous trap formers (STF) and endoparasites produced larger, or less volatile, slowly diffusing compounds, which attracted nematodes. However, the actual substances responsible for attraction remain unknown.

Nematodes are attracted to endozoic conidia which are adhesive, whereas ingested conidia do not attract nematodes. While the adhesive conidia can attach to any part of the nematode cuticle most infections occur in the head region, and it is within this area where the chemoreceptors are situated. Recent experiments by Jansson and Nordbring-Hertz[111] have shown that sialic acids located on the head region of nematodes are important in both chemotaxis and subsequent adhesion to the adhesive conidia produced by endoparasites such as *M. coniospora*. The nonattracting conidia of the *Harposporium* type endoparasites have strongly attracting mycelium, and it appears that nematodes are attracted to the mycelium where they unwittingly ingest the conidia along with other food material.[14]

D. Antibiotics

Fungal predators and parasites compete with other soil microorganisms, especially bacteria and fungi, for the available nutrients within the nematode body. However, nematophagous fungi have not been reported as being overwhelmed by bacteria and other fungi within the captured nematode. Although secondary microorganisms can enter moribund nematodes through any of the natural openings, such as the anus, vulva, excretory pore, or the buccal cavity, these competing organisms rarely develop. The assimilative or haustorial hyphae of many nematophagous Hyphomycetes release antibiotics within the nematode which are able to inhibit microorganisms.[4] This ability gives the group a definite competitive advantage over other soil microorganisms in the utilization of nematodes once captured or infected.

E. Trap Induction

In pure culture nematophagous fungi rarely produce trapping organs, but will produce traps abundantly within 24 hr if nematodes are added. The motility of the nematodes is a factor in trap induction, with slow-moving nematodes inducing a lower response than more active species.[127] Comandon and Foubrune[113] found that ring formation could be induced by adding sterile water in which nematodes had been living, suggesting that it was compounds secreted by the nematodes that induced trap formation. Since then numerous substances have been shown to induce trap formation, especially those of animal origin such as blood serum and earthworm extract. Substances which are able to induce trap formation are collectively known as "nemin".[130] Wootton and Pramer[131] established that the active agents in nemin were the amino acids valine, leucine, and isoleucine, with valine the most active. Nordbring-Hertz[132] confirmed that amino acids promoted trap formation in *A. oligospora* but found peptides and valyl peptides in particular more active. She suggested that the peptides capable of inducing trap formation in *A. oligospora* were produced by biological degradation of proteinaceous materials in soil. Nematodes themselves are much more effective in stimulating trap formation than either peptides or crude extracts from nematodes. Trap induction began within 2 hr and reached a maximum trap density within 5 hr, independent of nematode density.[133] So it appears that peptides and amino acids are only partially responsible for trap induction and the precise mechanism remains unknown, a conclusion supported by the work of Rosenzweig.[134] Although trap formation in *A. oligospora* requires induction by peptides or living nematodes, it is also dependent on an endogenous rhythm.[135] While temperature influenced hyphal extension it did not affect periodicity of trap formation even though at lower temperatures the peaks of trap formation were close enough together to give partial overlapping. Periodic trap formation at 21°C had a mean period length of 42.3 ± 0.8hr. Investigating how the presence of nematodes stimulates trap formation in *Arthrobotrys oviformis*, Cayrol et al.[136] concluded that trap induction possibly results from a complex relationship between the fungus and the nematode-associated bacteria. When stimulated the bacteria induce trap formation. They concede that the biochemical reactions have yet to be determined, but feel that there is a real possibility of increasing the nematophagous action of predatory Hyphomycetes by introducing appropriate bacteriophagous nematodes to stimulate bacterial activity.

Cooke[137] reminds us that species of nematophagous fungi differ in the degree of their response to nematode extracts and that different isolates of a single species can display a range of responses to such stimuli in terms of trap formation. He concludes that trap formation is not simply a response to exogenous stimuli, but a more complex response to the needs of the fungus.

VI. ECOLOGY

A. Distribution

There have been numerous surveys on the occurrence of nematophagous fungi which have shown that the group is found throughout the world and from all types of climates. The most notable surveys have been conducted in North America, in Maryland,[59,138] Oregon,[139] Florida, North Carolina,[140] California,[115] and Illinois;[141,142] and in Canada, in Quebec[143] and Ontario.[82] Surveys have also been conducted throughout Europe, in England,[144-150] Scotland,[151] Ireland,[9,152] Italy,[153] Poland,[154] Germany,[155] France,[66,136,156-158] USSR,[159-161] Denmark,[162] and Finland,[163,164] as well as in India,[32,33,165-169] Malaysia,[170] Japan,[171-174] Australia, in Queensland[175] and Western Australia,[176] New Zealand,[70,71] and the maritime Antarctic.[177,178] And individual new species are continuously being discovered and described from every corner of the world.[34,105,164,170,179-184]

The majority of species of nematophagous fungi appear fairly ubiquitous with few species restricted geographically. There is a steady stream of new species being isolated each year which of course appear to have a very limited distribution. However, this may be linked to scarcity and the limited amount of field data published. For example, the new species *Arthrobotrys pauca* isolated from Queensland in Australia by McCulloch in 1977[181] was subsequently isolated from Japan by Mitsui.[172] While isolates of some species have shown slight environmental adaptation, such as tolerance to lower temperatures in isolates from the Antarctic,[185] they are morphologically identical to isolates as far afield as Canada, Australia, India, and Europe. Gray[9] conducted a survey on the distribution and habitat of nematophagous fungi in Ireland, which he found resembled that reported for other temperate areas. He isolated 12 species of endoparasites and 20 species of predatory nematophagous fungi from 161 sites. Like Duddington[147] in his survey of British soils, *Acrostalagmus obovatus* and *Harposporium anguillulae* were the most frequently recorded endoparasites. Only one species was restricted to a single isolation, *Harposporium helicoides*, although Barron[82] found it to be fairly abundant in his survey of endoparasites from soils in Ontario. *Arthrobotrys oligospora* has generally been considered to be the most abundant predator in temperate soil. However Gray isolated *A. oligospora* from only 1.2% of his sites compared with *Dactylella bembicodes* from 8.7% of the sites. The most common predatory fungi from Ireland, *D. bembicodes*. *D. ellipsospora*, and *D. cionopaga*, are also amongst the most abundant predators found in British soil.[147] While the distribution of *D. mammillata* in Britain, *A. oligospora* in Ireland, and of *A. musiformis* in both Ireland and Britain appear highly restricted, all species are locally abundant.

B. Habitat

The most extensive field studies on the group have been carried out by Duddington,[147] Shepherd,[87] Fowler,[70] and McCulloch.[175] Although Duddington's study was nonquantitative and was restricted to only five habitat classifications, it remains the only detailed source of information on the habitat of nematophagous fungi from temperate soils, apart from occasional notes and observations. Gray[9] used ten broad terrestrial habitat classifications in his survey of Irish habitats (Table 2). He found that nematophagous fungi were abundant in all the habitats examined, although most widely in temporary agricultural pasture, coniferous leaf litter, and coastal vegetation, with over 90% of the samples containing nematophagous fungi. The greatest species diversity was recorded in deciduous leaf litter and permanent pasture. Although, when expressed as mean species diversity of infected plates, only peatland, dung, coniferous leaf litter, and coastal vegetation scored in excess of 2.0 species per plate. The semiquantitative method used in his survey and the small number of records for each species made the identification of associations between fungi and specific habitats difficult. Although *Nematoctonus leiosporus* and *Meria coniospora* were most frequently isolated from dung and compost, respectively, no endoparasite was restricted to any one habitat. *Acrostalagmus obovatus* was recorded from all the habitats except temporary agricultural pasture, compost, and peatland, suggesting it to be largely independent of habitat or soil conditions. Several distinct habitat associations were observed with the predatory fungi. *A musiformis* and *A. robusta*, both capturing nematodes by adhesive nets, were significantly more abundant in permanent pasture as was *D. cionopaga* which captures its prey with adhesive branches that often form two-dimensional nets. These similar trapping mechanisms were most frequently recorded from permanent pasture. *Dactylella lobata* was only isolated from deciduous leaf litter, as was *Stylopage hadra*, while *D. mammillata* was found equally abundantly in both deciduous and coniferous leaf litter. Some species appear adapted to specific habitats such as salt marshes;[186,187] and Mankau[1] has isolated some unusual species from specialized habitats such as arid desert soils and coastal sand dunes. A number of new species of predatory fungi (*Arthrobotrys constringens* and *Dactylella multiformis*) have been isolated from soil taken from a marsh at Manitoba in Canada.[182,183] It would

Table 2

NUMBER OF ISOLATES AND SPECIES OF NEMATOPHAGOUS FUNGI ISOLATED FROM EACH HABITAT SAMPLED BY GRAY,' IN HIS SURVEY OF IRISH SOILS, BY THEIR ATTACK MECHANISM

Endoparasites habitat (total)	Total fungi	Species diversity	No. of mean species sites (diversity of sampled infected plates)	% of sites with fungi	Predators						
					Adhesive hypha	Adhesive branches	Adhesive knobs	Constricting rings	Adhesive nets	Unknown	Total
Deciduous leaf litter 12	32	15	24 (1.78)	75	3	2	7	5	3	0	20
Coniferous leaf litter 10	24	7	12 (2.18)	92	0	0	8	5	0	1	14
Dung-old and partly revegetated 16	25	8	15 (2.50)	67	0	1	2	4	2	0	9
Permanent pasture 10	37	13	48 (1.42)	54	2	7	4	2	12	0	27
Temporary agricultural pasture 5	15	9	8 (1.88)	100	2	0	3	3	2	0	10
Cultivated land 4	10	6	12 (1.25)	67	1	0	0	2	3	0	6
Moss cushions 10	22	10	17 (1.69)	77	0	1	3	6	2	0	12
Decaying vegetation and compost 5	8	4	9 (1.60)	56	0	0	1	2	0	0	3
Peatland 7	9	5	4 (3.00)	75	0	0	1	1	0	0	2
Coastal vegetation 14	23	10	12 (2.09)	92	0	0	6	2	1	0	9
Totals	161				8	11	35	32	25	1	112

appear that nematophagous fungi may be present in a variety of unexplored habitats where nematodes abound, with many new species to discover. For example, the new species of *Arthrobotrys*, *A. ellipsospora*, isolated from pine sap below the bark and cambium layer in a red pine (*Pinus densiflora*) in Japan has already been discussed.[124]

C. Factors Affecting Distribution

The effect of the major soil variables such as soil moisture, organic matter, pH, nematode density, soil nutrients, and metals, on the distribution of nematophagous fungi has been extensively studied by Gray.[188-190] The endoparasitic nematophagous fungi are obligate parasites, and unlike the predatory fungi they are unable to live saprophytically in the soil. This is confirmed in temperate soils by all the endoparasitic groups being isolated from soil with significantly higher densities of nematodes compared with predatory groups. The mean density of nematodes in soil containing endoparasites which form nonattracting ingested conidia was much greater, compared with the soil samples from which attracting adhesive conidia producing species were isolated. This suggests that such a nonspecific method of attraction may rely on a greater density of soil nematodes to ensure infection compared with parasites which produce adhesive conidia. Those endoparasites producing conidia, both ingested and adhesive types, were however, significantly associated with soils with high organic matter, even though no correlation was established between organic matter and nematode density. Generally, the conidia-forming endoparasites were isolated from samples with comparatively high soil moisture contents and low pH. A major limitation of Gray's work was the use of moisture content of soil rather than water potential, especially as the types of soils used ranged from sandy loams to clay soils.[188] While water content may be important in terms of mobility of zoospores of the encysting endoparasites, a phenomenon also noted by Kerry et al.,[191] it is of much less value when assessing the effect of soil water on the distribution of soil fungi.[192]

Cooke[98] concluded that the development of predaceous efficiency had been accompanied by a tendency to lose those characters associated with an efficient saprophytic existence in the soil, namely rapid growth rate and good competitive saprophytic ability. This is supported by Gray[188] who found species producing adhesive networks are isolated significantly more frequently ($p<0.001$) from soils with low moisture and organic matter contents. In contrast, ring-forming species are isolated significantly more frequently ($p<0.001$) from soil with relatively high moisture and organic matter contents. None of the groups of species, with the exception of species forming adhesive branches, were associated with nematode density, however the latter conditions favor a more active microbial community with a greater biomass of organisms including nematodes. Under conditions of active microbial activity the property of spontaneous trap formation may confer an ecological advantage on the organism, especially if the fungus is sensitive to competition from other microorganisms. The poorer conditions of the soil which favor the net-forming species are the conditions where nonspontaneous trap formers would flourish. Their good saprophytic ability allows them to compete favorably with other species for the limited nutrients. Cooke[193] speculated that when conditions temporarily improve they may be able to maintain a competitive advantage by utilizing the subsequent increase in the nematode population.

Little is known, and few records exist, of the distribution of nematophagous fungi with unmodified adhesive hyphae. Although they are capable of very high levels of predaceous activity they are often observed without having caught any nematodes, even though potential prey may be abundant.[35] Gray's results indicate that they were isolated from soils with similar soil conditions as net-forming species.

Although the ring and knob-forming species are associated with soils of a low pH ($p<0.001$), those species with unmodified hyphae were isolated significantly more frequently from soils with a higher pH ($p<0.005$). Whether individual species have adapted to specific pH ranges,

Table 3
THE IMPORTANCE OF PARTICULAR SOIL FACTORS (INDEPENDENT VARIABLES) AS SELECTED BY STEPWISE LOGISTIC REGRESSION ON THE PRESENCE OF NEMATOPHAGOUS FUNGI CATEGORIZED BY TRAPPING MECHANISM OR MODE OF INFECTION (DEPENDENT VARIABLE), AS MEASURED IN IRISH SOILS[100]

Terms entered into model at each step (P<0.01)

Dependent variable	Step	Independent variable	Improvement in prediction (P-value)
Endoparasite	1	Organic matter	0.000
	2	pH	0.021
Encystment	1	Moisture	0.056
	2	Nematode density	0.095
Adhesive conidia	1	Organic matter	0.014
Predator	1	pH	0.00
Net	1	Moisture	0.000
Ring	1	Moisture	0.000
	2	pH	0.002
Unmodified hyphae	1	pH	0.037
Branch	0		
Knobs	1	pH	0.000
All groups	1	pH	0.001
	2	Organic matter	0.051
	3	Moisture	0.079

thus reducing interspecific competition, remains unclear, although the presence of many of the predatory groups of fungi is determined more by pH than any of the factors tested (Table 3).

The soil nutrients N, P, and K were found to be all positively correlated with nematode density.[90] Endoparasites with adhesive conidia are independent of soil nutrient, while those species with ingested conidia are isolated from soils with high concentrations of nutrients, indicating a strong dependence on a large nematode density. Knob-forming predators which rely on their ability to produce traps spontaneously are isolated from soils with low concentrations of nutrients, while those species with constricting rings are isolated from richer soils which contain a greater density of nematodes. Net-forming species are largely independent of soil fertility, although generally they are isolated from soils with limited nutrients, especially low K.

Heavy metals have been shown to have an adverse effect on microbial populations and on fungi in particular,[194,195] although some fungi are stimulated by low levels of metals.[196] Fungal species vary in their sensitivity to specific metals[194,197] and nematophagous fungi appear no different to other fungi in this respect. The natural concentrations of heavy metals in the soil examined by Gray[190] generally restricted the distribution of nematophagous fungi, with their presence being significantly associated with soils containing low concentrations of Cr (p <0.001), Ni (p<0.005), and Cu (p<0.05). Knob-forming predators were especially sensitive to Cu, Ni, and Zn, being recorded in soils with mean concentrations of 6.1, 12.9, and 25.4 μg/g respectively (Table 4). They were isolated exclusively from soil with low K, Ca, and Mg. Nematophagous fungi are affected by heavy metals in two ways. In the soil,

Table 4
THE EFFECT OF METAL CONCENTRATION IN
IRISH SOILS ON THE PRESENCE AND ABSENCE
OF KNOB-FORMING PREDATORS[190]

	Present (μg/g)			Absent (μg/g)			
	x	Sem	n	x	Sem	n	p
Pb	26.0	6.06	5	130.5	24.01	43	0.030
Zn	25.4	3.84	5	205.4	24.95	43	0.001
Cr	16.5	5.20	5	20.3	1.56	43	0.370
Ni	12.9	2.05	5	39.7	3.42	43	0.007
Cu	6.1	1.31	5	38.5	4.17	43	0.002
Ca	1567.9	720	5	30074.4	5430	43	0.005
Mg	1505.3	372	5	3932.6	296	43	0.003
K	835.5	108	5	3744.6	726	43	0.002

Note: The results include the mean (x), standard error of the mean (sem),
the level of significance of the Mann-Whitney statistic u using a normal
two-tail approximation (p), and the number of samples (n).

metals have a direct effect on NSTF living saprophytically or, accumulated in prey they have an indirect effect on endoparasites and STF. However, those species such as endoparasites with adhesive conidia, which are independent of the soil and attract prey directly, may have an advantage in this respect by completing life cycles using nutrients from the host only. They may be able to survive quite high concentrations of heavy metals in the soil so long as sufficient nematode hosts are present. This is assuming that the accumulated metals in the host do not inhibit either germination or growth of the infective mycelium within the captured organism. Such an endoparasite, *Meria coniospora*, was found to be tolerant to all metals and to Cu in particular. Interestingly, Rosenzweig and Pramer[198] have shown that collagenase production, which facilitates fungal penetration of the nematode cuticle, is less sensitive than either growth or trap formation to heavy metal inhibition. So it would appear that STFs would be less restricted by heavy metals than NSTFs. There is an increasing problem of contamination of agricultural soils by heavy metals, through deposition of agricultural practices such as chemical fertilization, dredging or sewage sludge disposal, pesticide application, and the particular problems of atmospheric dust fall and acidification.[199] For this reason a tolerance to heavy metals may be a prerequisite for the selection of suitable nematophagous fungal species for use as a biological control agent.

Little research has been done on the effects of pesticides on nematophagous fungi in the soil. If the group is to be used for the biological control of soil-borne nematodes, then they must be able to tolerate at least the residual concentrations found in most soils, and preferably be entirely compatible with the widely used pesticides at the concentrations normally used. Cayrol and Frankowski[200] examined the effect on growth of a range of herbicides on the fungus *A. irregularis*. They found that at concentrations of 10 ppm there was little or no effect on the fungus and even at 100 ppm some growth was still reported (Table 5). Bloxham[201] measured the response of another net-forming predator, *A. robusta*, to the fungicide Mancozeb. She found that the fungicide had a measurable inhibitory effect on germination, growth, sporulation, and nemin-induced trap formation. Sporulation and trap formation were particularly sensitive to the fungicide.

Predatory fungi found in the Antarctic are not truly psychrophilic, and although they are able to grow at significantly lower temperatures, they cannot do so rapidly.[185] The environmental conditions exerted on the microbial flora of Antarctic soil are very different to those in temperate soil. In the Antarctic the STF species are able to use internal resources only

Table 5
EFFECT OF HERBICIDES ON
THE GROWTH RATE OF
ARTHROBOTRYS
***OLIGOSPORA* WHICH IS**
SOLD COMMERCIALLY AS
ROYAL 350[200]

	Concentration	
Herbicide	**10 ppm**	**100 ppm**
Atrazine	100	100
Diquat	70	35
Glyphosate	100	25
Linuron	100	84
Metamitrone	100	100
Metribuzine	100	100
Monalide	72	36
Monolinuron	100	26
Paraquat	100	35
Penoxaline	78	45
Phenmediphame	100	90
Propachlore	100	67
Propyzamide	79	71
Simazine	100	100
Control	100	100

Note: The growth rate is the mean extension
of the mycelium after 1 day (mm).

and so are largely unaffected by soil conditions. They are able to respond to short periods of suitable weather when the soil is unfrozen and the temperature is well above 0°C, are able to attract nematodes, and only germinate and form traps when induced by the direct chemical stimulation of nematodes within the immediate vicinity of the conidium. Apart from endoparasites, the indigenous nematode density of the soils appears to be the least important soil variable in determining the presence of nematophagous fungi in temperate soils. In the Antartcic, the abiotic soil variables measured by Gray[189] appear to have little direct effect on distribution of the fungi, while the presence and abundance of prey appears to be of paramount importance. Predators are very strongly associated ($p < 0.001$) with soils containing a high nematode density.

D. Spatial Distribution
1. Vertical Distribution
The occurrence of certain species and groups of fungi are associated with specific soil variables and in particular pH, soil moisture, N, P, K, and soil density.[188,190] These soil variables are known to vary with depth, as are the densities of soil bacteria, fungi, and nematodes.[202] Peterson and Katznelson[77] recorded high levels of nematode-trapping activity from the rhizosphere area, and since then it is from this area that soil samples are normally taken during surveys of the group. The species of nematode-trapping fungi have been shown to vary with depth.[203] Mitsui et al.[204] found the greatest diversity in the upper 10 to 30 cm of soils, as well as a positive correlation between the population density of nematophagous fungi and root-knot nematodes in peanut fields. This confirmed an earlier study by Kobayashi and Mitsui[174] who correlated fungi with free-living nematodes in greenhouse soils. Gray and

Table 6
FREQUENCY OF OCCURRENCE OF
NEMATOPHAGOUS FUNGI BY TRAPPING
MECHANISM AND ABILITY TO
SPONTANEOUSLY PRODUCE TRAPS IN THE
HEMIEDAPHIC AND EUEDAPHIC ZONES OF
A DECIDUOUS WOODLAND SOIL[205]

	Hemiedaphic zone		Euedaphic zone	
	(n)	(%)	(n)	(%)
Endoaparsites	0	0.0	5	23.8
Adhesive branched spp.	2	9.5	0	0.0
Adhesive knob spp.	5	23.8	2	9.5
Constricting ring spp.	2	9.5	0	0.0
Spontaneous trap formers	9	42.9	2	9.5
Adhesive net spp.				
Nonspontaneous trap formers	1	4.8	4	19.1
All nematophagous fungi	10	47.6	11	52.4

Bailey[205] examined the vertical distribution of nematophagous fungi in soil cores collected from a deciduous woodland. Predators forming constricting rings, adhesive branches, and adhesive knobs are restricted to the upper litter and humus layer (hemiedaphic zone), while the net-forming predators and endoparasites are isolated at all depths, although they are significantly more abundant in the lower mineral rich soils (euedaphic zone). Like Mitsui et al.[204] they found a much greater species diversity of nematophagous fungi in the upper organic zones (Table 6). Predators able to form traps spontaneously are restricted to the organic soils of the hemiedaphic zone which are rich in nematodes. The reduced organic matter, moisture, and soil nutrients associated with the soil of the euedaphic zone, and with increasing depth, are conditions in which NSTF generally are able to flourish. Their good saprophytic ability allows them to compete favorably with other species for the limited nutrients available. Endoparasites are also found in the euedaphic zone, although they do not appear to compete for the very low density of prey organisms. The only endoparasite isolated by Gray and Bailey[205] was *Cephalosporium balanoides* which produces adhesive conidia which are able to chemically attract nematodes.[14] The competitive advantage that such endoparasites have over those producing ingested conidia is clearly seen. *C. balanoides* is isolated throughout the euedaphic zone which has a low nematode density. The results indicate that species of nematophagous fungi are indeed affected by depth, with the varying abiotic and biotic factors providing at least two distinct habitats. In a recent study of the group from Irish soils, samples were taken mainly from the rhizosphere area, and while *C. balanoides* was isolated, *Dactylaria psychrophila* and *Arthrobotrys pectospora* were not.[9] The latter two species are principally associated with moss cushions, while Gray and Bailey[205] isolated them from dry sandy soil in the euedaphic zone below the deciduous leaf litter. These habitats are similar in that they both undergo severe desiccation. This may indicate that *D. psychrophila* and *A. pectospora* are tolerant of low soil moisture,[189] which appears to be an important factor in their distribution.

2. Horizontal Distribution

Nematophagous fungi are small enough to be affected by microclimates within the soil. Since environmental conditions can vary quite dramatically locally, even over an area of a few centimeters, it is not surprising that horizontal studies based on small areas of a few square meters have shown considerable variation in species diversity. Four studies have been done, none of which have been extensive or subject to rigorous statistical analysis.[87,158,173,203] All have shown that specific species are not always recorded in each quadrat within the sample area, and neither is a species isolated on one occasion from a quadrat definitely isolated on subsequent occasions. For example, Shepherd,[87] working on Danish soils, found the net-forming species *Arthrobotrys oligospora* the most widely isolated species in her 1 m² sample area which was subdivided into 36 quadrats. On the three sampling occasions she isolated the species from 72, 83, and 86% of the quadrats. Both Peloille[158] and Mitsui[173] found *A. oligospora* to be most widely distributed, with Peloille recovering it each month over a 17 month sampling period in a permanent pasture grazed by sheep. She found that like *A. oligospora*, both *Arthrobotrys irregularis* and *Dactylaria candida* were evenly distributed throughout the pasture, and could be isolated from all the sites sampled. Mitsui[173] isolated five species in a peanut field. The two STF species *Arthrobotrys dactyloides* (constricting ring) and *Dactylella ellipsospora* (stalked adhesive knob) were not evenly distributed, being recorded far less frequently from fewer quadrats.

The ecology of nematophagous fungi and their role in the biological control of plant and animal parasitic nematodes has been reviewed elsewhere.[206]

REFERENCES

1. **Mankau, R.,** Biological control of nematode pests by natural enemies, *Annu. Rev. Phytopathol.*, 18, 415, 1980.
2. **Fresenius, G.,** *Beitr. Mykol.*, 1-2, 1, 1852.
3. **Zopf, W.,** Zur Kenntnis der Infektions — krankheiten niederer Thiere und Pflanzen, *Nova Acta Leop. Carol.*, 52, 314, 1888.
4. **Barron, G. L.,** The nematode – destroying fungi, in *Topics in Mycobiology*, Vol. 1, Canadian Biological Publications, Guelph, 1977, 140.
5. **Barron, G. L.,** Predators and parasites of microscopic animals, in *Biology of Conidial Fungi*, Vol. 2, Cole, G. T. and Kendryck, B., Eds., Academic Press, London, 1981, chap. 20.
6. **Barron, G. L.,** Nematode – destroying fungi, in *Experimental Microbial Ecology*, Burns, R. G. and Slater, J. H., Eds., Blackwell, Oxford, 1982, chap. 31.
7. **Peloille, M.,** Bibliographical review: nematode trapping fungi. Predatory phenomenon, ecology, use in biological control, *Agronomie*, 1, 331, 1981.
8. **Lysek, G. and Nordbring-Hertz, B.,** Die biologie Nematodenfangender Pilze, *Forum Mikrobiol.*, 6, 201, 1983.
9. **Gray, N. F.,** Ecology of nematophagous fungi: distribution and habitat, *Ann. Appl. Biol.*, 102, 501, 1983.
10. **Barron, G. L. and Percy, J. G.,** Nematophagous fungi: a new *Myzocytium*, *Can. J. Bot.*, 53, 1306, 1975.
11. **Davidson, J. G. N. and Barron, G. L.,** Nematophagous fungi: *Haptoglossa*, *Can. J. Bot.*, 51, 1317, 1973.
12. **Saikawa, M.,** Fixation of germinating conidia of *Meria coniospora* Drechsler with KMnO₄, *J. Electron Microsc.*, 31, 276, 1982.
13. **Saikawa, M.,** An electron microscope study of *Meria coniospora*, an endozoic nematophagous Hyphomycete, *Can. J. Bot.*, 60, 2019, 1982.
14. **Jansson, H. B.,** Attraction of nematodes to endoparasitic nematophagous fungi, *Trans. Br. Mycol. Soc.*, 79, 25, 1982.
15. **Dowsett, J. A., Reid, J., and Hopkin, A.,** On *Cephalosporium balanoides* Drechsler, *Mycologia*, 74, 687, 1982.
16. **Jansson, H. B. and Nordbring-Hertz, B.,** The endoparasitic fungus *Meria coniospora* infects nematodes specifically at the chemosensory organs, *J. Gen. Microbiol.*, 129, 1121, 1983.
17. **Jansson, H. B., Jeyaprakash, A., and Zuckerman, B. M.,** Differential adhesion and infection of nematodes by the endoparasitic fungus *Meria coniospora* (Deuteromycetes), *Appl. Environ. Microbiol.*, 49, 552, 1985.

18. **Saikawa, M., Totsuka, J., and Morikawa, C.**, An electron microscope study of initiation of infection by conidia of *Harposporium oxycoracum*, an endozoic nematophagous Hyphomycete, *Can. J. Bot.*, 61, 893, 1983.
19. **Saikawa, M.**, Ultrastructure of conidia of *Harposporium anguillulae* Lohde, *J. Electron Microsc.*, 31, 90, 1982.
20. **Drechsler, C.**, A Harposporium infecting eelworms by means of externally adhering awl-shaped conidia, *J. Wash. Acad. Sci.*, 40, 405, 1950.
21. **Saikawa, M. and Morikawa, C.**, An electron microscope study of initiation of infection by conidia of *Harposporium subuliforme*, an endozoic nematophagous fungus, *Trans. Mycol. Soc. Japan*, 26, 215, 1985.
22. **Dowsett, J. A. and Reid, J.**, Multilaminate bodies: additional candidates for membrane reserve in trapping rings of *Dactylaria brochopaga*, *Mycologia*, 75, 1094, 1983.
23. **Shepherd, A. M.**, Formation of the infection bulb in *Arthrobotrys oligospora* Fresenius, *Nature (London)*, 175, 475, 1955.
24. **Olthof, T. H. A. and Estey, R. H.**, A nematotoxin produced by the nematophagous fungus, *Arthrobotrys oligospora* Fresenius, *Nature (London)*, 196, 514, 1963.
25. **Balan, J. and Gerber, N. N.**, Attraction and killing of the nematode *Panagrellus redivivus* by the predacious fungus *Arthrobotrys dactyloides*, *Nematologica*, 18, 163, 1972.
26. **Giuma, A. Y., Hacket, A. M., and Cooke, R. C.**, Thermostable nematotoxins produced by germinating conidia of some endozoic fungi, *Trans. Br. Mycol. Soc.*, 60, 49, 1973.
27. **Krizkova, L., Balan, J., Nemec, P., and Kolozsvary, A.**, Predacious fungi *Dactylaria pyriformis* and *Dactylaria thaumasia* and nematicides, *Fol. Microbiol.*, 21, 493, 1976.
28. **Giuma, A. Y. and Cooke, R. C.**, Nematoxin production by *Nematoctonus haptocladus* and *N. concurrens*, *Trans. Br. Mycol. Soc.*, 56, 89, 1971.
29. **Barron, G. L.**, Nematophagous fungi: a new Arthrobotrys with nonseptate conidia, *Can. J. Bot.*, 57, 1371, 1979.
30. **Gray, N. F.**, *Monacrosporium psychrophilum*, a nematode-destroying fungus new to Ireland, *Ir. J. Agr. Res.*, 24, 129, 1985.
31. **Jansson, H. B. and Nordbring-Hertz, B.**, Trap and conidiophore formation in *Arthrobotrys superba*, *Trans. Br. Mycol. Soc.*, 77, 205, 1981.
32. **Sachchidanandia, J. and Swarup, G.**, Nematophagous fungi in Delhi soils, *Indian Phytopathol.*, 19, 279, 1966.
33. **Sachchidanandia, J. and Swarup, G.**, Additional nematophagous fungi from Delhi soils, *Curr. Sci.*, 36, 677, 1967.
34. **Wood, S. N.**, *Stylopage anomala* sp. nov. from dung, *Trans. Br. Mycol. Soc.*, 80, 368, 1983.
35. **Gray, N. F.**, Ecology of nematophagous fungi: predatory and endoparasitic species new to Ireland, *Ir. Nat. J.*, 21, 337, 1984.
36. **Bucaro, R. D.**, *Hongos nematofagos* de El Salvador, *Rev. Biol. Trop.*, 31, 25, 1983.
37. **Dowsett, J. A., Reid, J., and Hopkin, A. A.**, Microscopic observations on the trapping of nematodes by the predaceous fungus *Dactylella cionopaga*, *Can. J. Bot.*, 62, 674, 1984.
38. **Gray, N. F.**, The effect of fungal parasitism and predation on the population dynamics of nematodes in the activated sludge process, *Ann. Appl. Biol.*, 104, 143, 1984.
39. **Nordbring-Hertz, B. and Stalhammar-Carlemalm, M.**, Capture of nematodes by *Arthrobotrys oligospora*, an electron microscope study, *Can. J. Bot.*, 56, 1297, 1978.
40. **Veenhuis, M., Nordbring-Hertz, B., and Harder, W.**, Occurrence, characterization and development of two different types of microbodies in the nematophagous fungus Arthrobotrys oligospora, *FEMS Microbiol. Lett.*, 24, 31, 1984.
41. **Dowsett, J. A. and Reid, J.**, Light microscope observations on the trapping of nematodes by *Dactylaria candida*, *Can. J. Bot.*, 55, 2956, 1977.
42. **Dowsett, J. A. and Reid, J.**, Transmission and scanning electron microscope observations on the trapping of nematodes by *Dactylaria candida*, *Can. J. Bot.*, 55, 2963, 1977.
43. **Wimble, D. B. and Young, T. W. K.**, Structure of adhesive knobs in *Dactylella lysipaga*, *Trans. Br. Mycol. Soc.*, 80, 515, 1983.
44. **Wimble, D. B. and Young, T. W. K.**, Ultrastructure of the infection of nematodes by *Dactylella lysipaga*, *Nova Hedwigia*, 40, 9, 1984.
45. **Schenck, S., Chase, T., Rosenzweig, W. D., and Pramer, D.**, Collogenase production by nematode trapping fungi, *Appl. Environ. Microbiol.*, 40, 567, 1980.
46. **Estey, R. H. and Tzean, S. S.**, Scanning electron microscopy of fungal nematode-trapping devices, *Trans. Br. Mycol. Soc.*, 66, 520, 1976.
47. **Barron, G. L.**, Detachable adhesive knobs in *Dactylaria*, *Trans. Br. Mycol. Soc.*, 65, 311, 1975.
48. **Saikawa, M.**, Ultrastructural features of the non-constricting ring trap in *Dactylella leptospora*, *Trans. Mycol. Soc. Japan*, 26, 209, 1985.

49. **Dowsett, J. A. and Reid, J.**, Observations on the trapping of nematodes by *Dactylaria scaphoides* using optical, transmission and scanning-electron microscope techniques, *Mycologia*, 71, 379, 1979.

50. **Insell, J. P. and Zachariah, K.**, A biometrical analysis of the giant constricting ring mutant of the predacious fungus *Dactylella brochopaga*, *Protoplasma*, 93, 305, 1977.

51. **Insell, J. P. and Zachariah, K.**, Some ring-trap mutants of the fungus *Dactylella brochopaga* Drechsler, *Arch. Microbiol.*, 117, 221, 1978.

52. **Couch, J. N.**, The formation and operation of the traps in the nematode catching fungus *Dactylella bembicodes* Drechsler, *J. Elisha Mitchell Sci. Soc.*, 53, 301, 1937.

53. **Muller, H. G.**, The constricting ring mechanism of two predaceous Hyphomycetes, *Trans. Br. Mycol. Soc.*, 41, 341, 1958.

54. **Lawton, J. R.**, The formation and closure of constricting rings in two nematode-catching Hyphomycetes, *Trans. Br. Mycol. Soc.*, 50, 195, 1967.

55. **Heintz, C. E. and Pramer, D.**, Ultrastructure of nematode-trapping fungi, *J. Bacteriol.*, 110, 1163, 1972.

56. **Rudek, W. T.**, The constriction of the trapping rings in *Dactylaria brochopaga*, *Mycopathologia*, 55, 193, 1975.

57. **Dowsett, J. A., Reid, J., and Caeseele, L., van**, Transmission and scanning electron microscope observations on the trapping of nematodes by *Dactylaria brochopaga*, *Can. J. Bot.*, 55, 2945, 1977.

58. **Insell, J. P. and Zachariah, K.**, The mechanism of the ring trap of the predacious Hyphomycete *Dactylella brochopaga* Drechsler, *Protoplasma*, 95, 175, 1978.

59. **Drechsler, C.**, Several species of Dactylella and Dactylaria that capture free-living nematodes, *Mycologia*, 42, 1, 1950.

60. **Higgins, M. L. and Pramer, D.**, Fungal morphogenesis: ring formation and closure by *Arthrobotrys dactyloides*, *Science*, 155, 345, 1967.

61. **Tzean, S. S. and Estey, R. H.**, Transmission electron microscopy of fungal nematode-trapping devices, *Can. J. Plant Sci.*, 59, 785, 1979.

62. **Dollfus, R. P.**, *Parasites (animaux et vegetaux) des Helminthes*, Paul Lechevalier, Paris, 1946, 481.

63. **Cooke, R. C. and Godfrey, B. E. S.**, A key to the nematode-destroying fungi, *Trans. Br. Mycol. Soc.*, 47, 61, 1964.

64. **Gray, N. F.**, Ecology of nematophagous fungi: notes on the identification of *Acrostalagmus obovatus* Drechsler, *Cephalosporium balanoides* Drechsler and *Verticillium sphaerosporum* Goodey, *Ir. Nat. J.*, 21, 125, 1983.

65. **Barron, G. L.**, Nematophagous fungi: Protascus and its relationship to Myzocytium, *Can. J. Bot.*, 55, 819, 1977.

66. **Peloille, M. and Cayrol, J. C.**, Premier isolement en France de deux especes d'Hyphomycetes predateurs de Nematodes: Arthrobotrys oviformis Sop. and Arthrobotrys conoides Drech, *Rev. Mycol.*, 43, 219, 1979.

67. **Gams, W. and Jansson, H. B.**, The nematode parasite *Meria coniospora* Drechsler in pure culture and its classification, *Mycotaxon*, 22, 33, 1985.

68. **Duddington, C. L. and Wyborn, C. H. E.**, Recent research on the nematophagous hyphomycetes, *Bot. Rev.*, 38, 545, 1972.

69. **Gams, W. and Zaayen, A. van**, Contribution to the taxonomy and pathogenicity of fungicolous Verticillium species. I. Taxonomy, *Neth. J. Plant Pathol.*, 88, 57, 1982.

70. **Fowler, M.**, New Zealand predacious fungi, *N. Z. J. Bot.*, 8, 283, 1970.

71. **Wood, F. H.**, Nematode-trapping fungi from a tussock grassland soil in New Zealand, *N. Z. J. Bot.*, 11, 231, 1973.

72. **Schenck, S., Kendrick, W. B., and Pramer, D.**, A new nematode-trapping Hyphomycete and a re-evaluation of Dactylaria and Arthrobotrys, *Can. J. Bot.*, 55, 977, 1977.

73. **Cooke, R. C. and Dickinson, C. H.**, Nematode-trapping species of Dactylella and Monacrosporium, *Trans. Br. Mycol. Soc.*, 48, 621, 1965.

74. **Tribe, H. T.**, Prospects for the biological control of plant-parasitic nematodes, *Parasitology*, 81, 619, 1980.

75. **Mankau, R.**, A semiquantitative method for enumerating and observing parasites and predators of soil nematodes, *J. Nematol.*, 7, 119, 1975.

76. **Peterson, E. A. and Katznelson, H.**, Occurrence of nematode-trapping fungi in the rhizosphere, *Nature (London)*, 204, 111, 1964.

77. **Peterson, E. A. and Katznelson, H.**, Studies on the relationships between nematodes and other soil microorganisms. IV. Incidence of nematode-trapping fungi in the vicinity of plant roots, *Can. J. Microbiol.*, 11, 491, 1965.

78. **Gray, N. F.**, Ecology of nematophagous fungi: methods of collection, isolation and maintenance of predatory and endoparasitic fungi, *Mycopathologia*, 86, 143, 1984.

79. **Wyborn, C. H. E., Priest, D., and Duddington, C. L.**, Selective technique for the determination of nematophagous fungi in soils, *Soil Biol. Biochem.*, 1, 101, 1969.

80. **Gray, N. F.,** Ecology of nematophagous fungi: *Panagrellus redivivus* as the target nematode, *Plant Soil*, 73, 293, 1983.
81. **Heintz, C. E.,** Assessing the predacity of nematode-trapping fungi *in vitro*, *Mycologia*, 70, 1086, 1978.
82. **Barron, G. L.,** Nematophagous fungi: endoparasites in Ontario and their ability to parasitize *Rhabditis terricola*, *Microb. Ecol.*, 4, 157, 1978.
83. **Dougherty, E.,** Cultivation of Ashelminthes, especially Rhabditis nematodes, in *Nematology*, Sasser, J. N. and Jenkins, W. R., Eds., University of North Carolina Press, Chapel Hill, N.C., 1960, chap. 6.
84. **Whitehead, A. G. and Hemmings, J. R.,** A comparison of some quantitative methods of extracting small vermiform nematodes from soil, *Ann. Appl. Biol.*, 55, 25, 1965.
85. **Gray, N. F.,** Ecology of nematophagous fungi: comparison of the soil sprinkling method with the Baermann funnel technique in the isolation of endoparasites, *Soil Biol. Biochem.*, 16, 81, 1984.
86. **Peters, B. G.,** A note on simple methods of recovering nematodes from soil, in *Soil Zoology*, Keven, D. K., Ed., Proc. Univ. Nottingham, 2nd Easter School of Agricultural Science, Butterworths, London, 1955, 373.
87. **Shepherd, A. M.,** Some observations on the distribution and biology of fungi predaceous on nematodes, Ph.D. thesis, University of London, 1955.
88. **Dixon, S. M. and Duddington, C. L.,** Permanent preparations of fungi growing on agar, *Nature (London)*, 168, 38, 1951.
89. **Faust, M. A. and Pramer, D.,** A staining technique for the examination of nematode-trapping fungi, *Nature (London)*, 204, 94, 1964.
90. **Aschner, M. and Kohn, S.,** The biology of *Harposporium anguillulae*, *J. Gen. Microbiol.*, 19, 182, 1958.
91. **Giuma, A. Y. and Cooke, R. C.,** Some endozoic parasites on soil nematodes, *Trans. Br. Mycol. Soc.*, 59, 213, 1972.
92. **Tarjan, A. C.,** Predaceous activity and growth of nematophagous fungi on various organic substances, *Phytopathology*, 50, 577, 1960.
93. **Coscarelli, V. and Pramer, D.,** Nutrition and growth of *Arthrobotrys conoides*, *J. Bacteriol.*, 84, 60, 1962.
94. **Blackburn, F. and Hayes, W. A.,** Studies on the nutrition of *Arthrobotrys oligospora* Fres. and *A. robusta* Dudd. I. The saprophytic phase, *Ann. Appl. Biol.*, 58, 43, 1966.
95. **Satchuthananthavale, V. and Cooke, R. C.,** Carbohydrate nutrition of some nematode-trapping fungi, *Nature (London)*, 214, 321, 1967.
96. **Satchuthananthavale, V. and Cooke, R. C.,** Vitamin requirements of some nematode-trapping fungi, *Trans. Br. Mycol. Soc.*, 50, 221, 1967.
97. **Satchuthananthavale, V. and Cooke, R. C.** Nitrogen nutrition of some nematode-trapping fungi, *Trans. Br. Mycol. Soc.*, 50, 423, 1967.
98. **Cooke, R. C.,** Ecological characteristics of nematode-trapping Hyphomycetes. I. Preliminary studies, *Ann. Appl. Biol.*, 52, 431, 1963.
99. **Zachariah, K. and Victor, J. R.,** A natural auxotroph of a nematode-trapping fungus, *Can. J. Bot.*, 61, 3255, 1983.
100. **Bricklebank, J. and Cooke, R. C.,** Utilisation of polysaccharides by two nematode-parasitic fungi, *Trans. Br. Mycol. Soc.*, 52, 347, 1969.
101. **Karling, J. S.,** A saprophytic species of Catenaria isolated from roots of *Panicum variegatum*, *Mycologia*, 26, 528, 1934.
102. **Birchfield, W.,** A new species of Catenaria parasitic on nematodes of sugar cane, *Mycopathologia*, 13, 331, 1960.
103. **Esser, R. P. and Ridings, W. H.,** Pathogenicity of selected nematodes by *Catenaria anguillulae*, *Soil Crop Sci. Fla.*, 33, 60, 1973.
104. **Jaffee, B. A.,** Parasitism of *Xiphinema rivesi* and *X. americanum* by zoosporic fungi, *J. Nematol.*, 18, 87, 1986.
105. **Sturhan, D. and Schneider, R.,** *Hirsutella heteroderae*, ein neuer nematodparasitarer Pilz., *Phytopathol. Z.*, 99, 105, 1980.
106. **Jaffee, B. A. and Zehr, E. I.,** Parasitism of the nematode *Criconemella xenoplax* by the fungus *Hirustella rhossiliensis*, *Phyopathology*, 72, 1378, 1982.
107. **Jaffee, B. A. and Zehr, E. I.** Parasitic and saprophytic abilities of the nematode-attacking fungus *Hirsutella rhossiliensis*, *J. Nematol.*, 17, 341, 1985.
108. **Capstick, C. K., Twinn, D. C., and Waid, J. S.,** Predation of natural populations of free-living nematodes by fungi, *Nematologica*, 2, 193, 1957.
109. **Pandey, V. S.,** Predatory activity of nematode trapping fungi against the larvae of *Trichostrongylus axei* and *Ostertagia ostertagi*: a possible method of biological control, *J. Helminthol.*, 47, 35, 1973.
110. **Jansson, H. B. and Nordbring-Hertz, B.,** Attraction of nematodes to living mycelium of nematophagous fungi, *J. Gen. Microbiol.*, 112, 89, 1979.

111. **Jansson, H. B. and Nordbring-Hertz, B.**, Involvement of sialic acid in nematode chemotaxis and infection by an endoparasitic nematophagous fungus, *J. Gen. Microbiol.*, 130, 39, 1984.

112. **Jansson, H. B.**, Predacity by nematophagous fungi and its relation to the attraction of nematodes, *Microbiol. Ecol.*, 8, 233, 1982.

113. **Comandon, J. and de Fonbrune, P.**, Recherches experimentales sur les champignons predateurs de nematodes du sol, *C. R. S. Soc. Biol., Paris*, 129, 619, 1938.

114. **Deschiens, R., Lamy, L., and Vautrin, E.**, Essai de pratiques de prophylaxie de l'anguillulose des vegetaux par l'emploi d'Hyphomycetes predateurs, *C. R. Seances Acad. Sci. Paris*, 216, 539, 1943.

115. **Mankau, R. and Clarke, O. F.**, Nematode-trapping fungi in Southern Californian citrus soils, *Plant Dis. Rep.*, 9, 968, 1959.

116. **Virat, M. and Peloille, M.**, Pouvoir predateur *in vitro* d'une souche d'Arthrobotrys oligospora Fres. vis a vis d'n nematode zooparasite, *Ann. Rech. Vet.*, 8, 51, 1977.

117. **Grønvold, J., Korsholm, H., Wolstrup, J., Nansen, P., and Henriksen, S. A.**, Laboratory experiments to evaluate the ability of *Arthrobotrys oligospora* to destroy infective larvae of *Cooperia* species, and to investigate the effect of physical factors on the growth of the fungus, *J. Helminthol.*, 59, 119, 1985.

118. **Rosenzweig, W. D., Premachandran, D., and Pramer, D.**, Role of trap lectins in the specificity of nematode capture by fungi, *Can. J. Microbiol.*, 31, 693, 1985.

119. **Nordbring-Hertz, B. and Mattiasson, B.**, Action of a nematode-trapping fungus shows lectin mediated host-microorganism interaction, *Nature (London)*, 281, 477, 1979.

120. **Nordbring-Hertz, B., Friman, E., and Mattiasson, B.**, A recognition mechanism in the adhesion of nematodes to nematode-trapping fungi, in *Lectins-Biology, Biochemistry and Clinical Biochemistry*, Vol. 2, Bog-Hansen, T. C., Ed., W. de Gruyter, Berlin, 1982, 83.

121. **Rosenzweig, W. D. and Ackroyd, D.**, Binding characteristics of lectins involved in the trapping of nematodes by fungi, *Appl. Environ. Microbiol.*, 46, 1093, 1983.

122. **Nordbring-Hertz, B.**, Mycelial development and lectin-carbohydrate interactions in nematode-trapping fungi, in *Ecology and Physiology of the Fungal Mycelium*, Jennings, D. H. and Rayner, A. D. M., Eds., Cambridge University Press, Cambridge, 1984, 419.

123. **Rosenzweig, W. D. and Ackroyd, D.**, Influence of soil micro-organisms on the trapping of nematodes by nematophagous fungi, *Can. J. Microbiol.*, 30, 1437, 1984.

124. **Tubaki, K. and Yamanaka, K.**, An undescribed nematode-trapping species of Arthrobotrys, *Trans. Mycol. Soc. Japan*, 25, 349, 1984.

125. **Balan, J., Krizkova, L., Nemec, P., and Vollek, V.**, Production of nematode-attracting and nematicidal substances by predacious fungi, *Folia Microbiol.*, 19, 512, 1974.

126. **Field, J. I. and Webster, J.**, Traps of predacious fungi attract nematodes, *Trans. Br. Mycol. Soc.*, 68, 467, 1977.

127. **Jansson, H. B. and Nordbring-Hertz, B.**, Interactions between nematophagous fungi and plant-parasitic nematodes: attraction, induction of trap formation and capture, *Nematologica*, 26, 383, 1980.

128. **Feder, W. A.**, A comparison of nematode-capturing efficiencies of five *Dactylella* species at four temperatures, *Mycopathol. Mycol. Appl.*, 19, 99, 1963.

129. **Monoson, H. L.**, Trapping effectiveness of five species of nematophagous fungi cultured with myceliophagous nematodes, *Mycologia*, 60, 788, 1968.

130. **Pramer, D. and Stoll, N. R.**, Nemin: a morphogenic substance causing trap formation by predaceous fungi, *Science*, 129, 966, 1959.

131. **Wootton, L. M. O. and Pramer, D.**, Valine induced morphogenesis in *Arthrobotrys coniodes*, *Bacteriol. Proc.*, 75, 1966.

132. **Nordbring-Hertz, B.**, Peptide-induced morphogenesis in the nematode trapping fungus *Arthrobotrys oligospora*, *Physiol. Plant.*, 29, 223, 1973.

133. **Nordbring-Hertz, B.**, Nematode-induced morphogenesis in the predacious fungus *Arthrobotrys oligospora*, *Nematologica*, 23, 443, 1977.

134. **Rosenzweig, W. D.**, Role of amino acids, peptides, and medium composition in trap formation by nematode-trapping fungi, *Can. J. Microbiol.*, 30, 265, 1984.

135. **Lysek, G. and Nordbring-Hertz, B.**, An endogenous rhythm of trap formation in the nematophagous fungus *Arthrobotrys oligospora*, *Planta*, 152, 50, 1981.

136. **Cayrol, J-C., Combettes, S., and Quiles, C.**, Influence de l'association nematodes-bacteries sur la formation des pieges chez l'Hyphomycete predateur *Arthrobotrys oviformis*, *Cryptogamie Mycologie*, 5, 21, 1984.

137. **Cooke, R. C.**, *The Biology of Symbiotic Fungi*, John Wiley & Sons, New York, 1977, 282.

138. **Drechsler, C.**, Some hyphomycetes that prey on free-living terricolous nematodes, *Mycologia*, 29, 447, 1937.

139. **Tolmsoff, W. J.**, The isolation of nematode trapping fungi from Oregon soils, *Phytopathology*, 49, 113, 1959.

140. **Feder, W. A.**, Nematophagous fungi recovered around Highlands, North Carolina, *Plant Dis. Rep.*, 46, 872, 1962.
141. **Monoson, H. L. and Williams, S. A.**, Endoparasitic nematode-trapping fungi of Mason State forest, *Mycopathol. Mycol. Appl.*, 49, 177, 1973.
142. **Monoson, H. L., Conway, T. D., and Nelson, R. E.**, Four endoparasitic nematode-destroying fungi from Sand Ridge State Forest soil, *Mycopathologia*, 57, 59, 1975.
143. **Estey, R. H. and Olthof, Th. H. A.**, The occurrence of nematophagous fungi in Quebec, *Phytoprotection*, 46, 14, 1965.
144. **Duddington, C. L.**, Predacious fungi from Cotswold leaf mould, *Nature (London)*, 144, 150, 1940.
145. **Duddington, C. L.**, Predacious fungi in Britain, *Trans. Br. Mycol. Soc.*, 29, 170, 1946.
146. **Duddington, C. L.**, Further records of British predacious fungi. I, *Trans. Br. Mycol. Soc.*, 33, 209, 1950.
147. **Duddington, C. L.**, The ecology of predacious fungi. I. Preliminary survey, *Trans. Br. Mycol. Soc.*, 34, 322, 1951.
148. **Duddington, C. L.**, Further records of British predaceous fungi. II, *Trans. Br. Mycol. Soc.*, 34, 194, 1951.
149. **Duddington, C. L.**, Nematode-destroying fungin in agricultural soil, *Nature (London)*, 173, 500, 1954.
150. **Juniper, A. J.**, Dung as a source of predacious fungi, *Trans. Br. Mycol. Soc.*, 40, 346, 1957.
151. **Mackenzie, D. W. R.**, Predacious fungi in the Edinburgh district, *Trans. Proc. Bot. Soc. Edinburgh*, 39, 35, 1960.
152. **Gray, N. F.**, The distribution of nematophagous fungi in Ireland, *Bull. Ir. Biogeog. Soc.*, 7, 19, 1984.
153. **Verona, O. and Lepidi, A. A.**, Primi reperti sulla presenza di micromiceti predatori di nematodi in alcuni terreni italiani, *Agri. Ital.*, 70, 217, 1970.
154. **Jarowaja, N.**, Preliminary investigations on fungi natural enemies of eelworms, *Biul. Inst. Ochr. Rosl.*, 21, 189, 1963.
155. **Fritsch, G. and Lysek, G.**, Nematodenzerstorende (endoparasitische) Pilze aus Waldboden yon der Pfaueninsel in Berlin, *Z. Mykol.*, 49, 183, 1983.
156. **Virat, M.**, Sur deux Hyphomycetes predateurs de nematodes isoles de praire et notes sur les genres Candelabrella et Duddingtonia, *Rev. Mycol.*, 41, 415, 1977.
157. **Peloille, M.**, Hyphomycetes predateurs de nematodes dans prairie du Limousin, *Entomophaga*, 26, 91, 1981.
158. **Peloille, M.**, Etude des Hyphomycetes predateurs de Nematodes rencontres sur une prairie du Limousin: Morphologie — physiologie — frequence et distribution, *These de l'Universite de Rennes*, 1981, 106.
159. **Soprunov, F. F. and Galiulina, Z. A.**, Predaceous hyphomycetes from Turkmenistan soil, *Mikrobiologiya*, 20, 489, 1951.
160. **Soprunov, F. F.**, *Predacious Hyphomycete Fungi and their Utilisation in the Control of Pathogenic Nematodes*, Golovin, P. N., Ed., Academic Sci. Turkmen. S.S.R. Ashabed, (In Russian), 1958, Translated — *Israel Program for Science Translations for U.S. Dept. Agriculture and the National Science Foundation*, Washington, D.C., 1966.
161. **Mekhtieva, N. A.**, Main results and outlooks of studying predacious fungi in Azerbaidjan, *Mikol. Fitopat.*, 6, 477, 1972.
162. **Shepherd, A. M.**, A short survey of Danish nematophagous fungi, *Friesia*, 5, 396, 1956.
163. **Ruokola, A-L. and Salonen, A.**, On nematode-destroying fungi in Finland, *Maataloust. Aikakausk.*, 39, 119, 1967.
164. **Salonen, A. and Ruokola, A. L.**, On nematode-destroying fungi in Finland. II, *Maataloust. Aikakousk.*, 40, 142, 1968.
165. **Das-Gupta, Shome, U. and Shome, K.**, Preliminary report on predacious fungi in India, *Curr. Sci.*, 33, 380, 1964.
166. **Dayal, R. and Nand, R.**, Studies in predaceous fungi. I. Zoopagales, *Proc. Nat. Acad. Sci., India*, 43(B), 87, 1973.
167. **Dayal, R. and Gupta, R. N.**, Studies in predacious fungi. IV. Some members of Hyphomycetes, *Proc. Nat. Acad. Sci., India*, 45(B), 237, 1975.
168. **Dayal, R. and Singh, V. K.**, Studies in predacious fungi. III. Some members of Hyphomycetes, *Proc. Nat. Acad. Sci., India*, 45(B), 89, 1975.
169. **Dayal, R. and Srivastava, S. S.**, Studies in predacious fungi. V. Some members of Hyphomycetes, *Proc. Nat. Acad. Sci., India*, 48(B), 88, 1978.
170. **Kuthubutheen, A. J., Muid, S., and Webster, J.**, Arthrobotrys dendroides, a new nematode-trapping synnematous *Arthrobotrys* from Malaysia, *Trans. Br. Mycol. Soc.*, 84, 563, 1985.
171. **Miura, K.**, Three entomophthoralean parasites of nematodes collected from Japan, *Rept. Tottori Mycol. Inst. (Japan)*, 10, 517, 1973.
172. **Mitsui, Y.**, The distribution, physiological and ecological nature of the nematode trapping fungi in Japan, *Bull. Nat. Inst. Agri. Sci. Ser. C*, 37, 127, 1983.

173. **Mitsui, Y.,** Distribution and ecology of nematode-trapping fungi in Japan, *JARQ,* 18, 182, 1985.

174. **Kobayashi, Y. and Mitsui, Y.,** Distribution of nematode-trapping fungi and their effect on the nematode population in soil, *Shizuoka Agric. Exp. Sta.,* 20, 41, 1975.

175. **McCulloch, J. S.,** A survey of nematophagous fungi in Queensland, *Queensl. J. Agric. Anim. Sci.,* 34, 25, 1977.

176. **Tan-Han-Kwang, B. S.,** Nematophagous fungi from Western Australia, M.Sc. thesis, University of Western Australia, Perth, Australia, 1966, 249.

177. **Gray, N. F., Wyborn, C. H. E., and Smith, R. I. L.,** Nematophagous fungi from the maritime Antarctic, *Oikos,* 38, 194, 1982.

178. **Gray, N. F. and Smith, R. I. L.,** The distribution of nematophagous fungi in the maritime Antarctic, *Mycopathologia,* 85, 81, 1984.

179. **Cooke, R. C. and Sachuthananthavale, V.,** Some nematode-trapping species of Dactylaria, *Trans. Br. Mycol. Soc.,* 49, 27, 1965.

180. **Monoson, H. L. and Pikul, F. J.,** A new endoparasitic species of Harposporium isolated from Illinois soil, *Mycologia,* 66, 178, 1974.

181. **McCulloch, J. S.,** New species of nematophagous fungi from Queensland, *Trans. Br. Mycol. Soc.,* 68, 173, 1977.

182. **Dowsett, J. A., Reid, J., and Kalkat, R. S.,** A new species of *Arthrobotrys* from soil, *Mycologia,* 76, 559, 1984.

183. **Dowsett, J. A., Reid, J., and Kalkat, R. S.,** A new species of *Dactylella, Mycologia,* 76, 563, 1984.

184. **Barron, G. L.,** A new Gonimochaete with an oospore state, *Mycologia,* 77, 17, 1985.

185. **Gray, N. F.,** Psychro-tolerant nematophagous fungi from the maritime Antarctic, *Plant Soil,* 64, 431, 1982.

186. **Jones, F. R.,** Three zoopagales from brackish water, *Nature (London),* 181, 575, 1958.

187. **Johnson, T. W. and Autery, C. L.,** An Arthrobotrys from brackish water, *Mycologia,* 53, 432, 1961.

188. **Gray, N. F.,** Ecology of nematophagous fungi: effect of soil moisture, organic matter, pH, and nematode density on distribution, *Soil Biol. Biochem.,* 17, 499, 1985.

189. **Gray, N. F.,** Nematophagous fungi from the maritime Antarctic: factors affecting distribution, *Mycopathologia,* 90, 165, 1985.

190. **Gray, N. F.,** Ecology of nematophagous fungi: effect of the soil nutrients N, P and K, and seven major metals on distribution, *Plant Soil,* in press.

191. **Kerry, B. R., Crump, D. H., and Mullen, L. A.,** Natural control of the cereal cyst nematode, *Heterodera avenae* Woll., by soil fungi at three sites, *Plant Prot.,* 1, 99, 1982.

192. **Griffin, D. M.,** *Ecology of Soil Fungi,* Chapman and Hall, London, 1972, 193.

193. **Cooke, R. C.,** Succession of nematophagous fungi during the decomposition of organic matter in soil, *Nature (London),* 197, 205, 1963.

194. **Babich, H. and Stotzky, G.,** Sensitivity of various bacteria, including actinomycetes, and fungi to cadmium and the influence of pH on sensitivity, *Appl. Environ. Microbiol.,* 33, 681, 1977.

195. **Bisessar, S.,** Effects of heavy metals on micro-organisms in solids near a secondary lead smelter, *Wat. Air and Soil Poll.,* 17, 305, 1982.

196. **Englander, C. M. and Corden, M. E.,** Stimulation of mycelial growth of *Endothia parasitica* by heavy metals, *Appl. Microbiol.,* 22, 1012, 1971.

197. **Babich, H. and Stotzky, G.,** Abiotic factors affecting the toxicity of lead to fungi, *Appl. Environ. Microbiol.,* 38, 506, 1979.

198. **Rosenzweig, W. D. and Pramer, D.,** Influence of cadmium, zinc, and lead on growth, trap formation and collagenase activity of nematode-trapping fungi, *Appl. Environ. Microbiol.,* 40, 694, 1980.

199. **Navarre, F. L., Ronneau, C., and Priest, P.,** Deposition of heavy metals on Belgian agricultural soils, *Wat. Air and Soil Poll.,* 14, 207, 1980.

200. **Cayrol, J-C. and Frankowski, J-P.,** Effet de quelques substances herbicides communes sur le champignon nematophage 'R350', *P. H. M. Rev. Hortic.,* 219, 52, 1981.

201. **Bloxham, A.,** The *in vitro* response of the predaceous fungus *Arthrobotrys robusta* to fungicide (mancozeb), B. A. thesis, University of Dublin, Dublin, Ireland, 1984, 53.

202. **Burges, A. and Raw, F.,** *Soil Biology,* Academic Press, London, 1967, 532.

203. **Mankau, R. and McKenny, M. V.,** Spatial distribution of nematophagous fungi associated with *Meloidogyne incognita* on peach, *J. Nematol.,* 8, 294, 1976.

204. **Mitsui, Y., Yoshida, T., Okamoto, K., and Ishii, R.,** Relationship between nematode trapping fungi and *Meloidogyne hapla* in the peanut field, *Jap. J. Nematol.,* 6, 47, 1976.

205. **Gray, N. F. and Bailey, F.,** Ecology of nematophagous fungi: vertical distribution in a deciduous woodland, *Plant Soil,* 86, 217, 1985.

206. **Gray, N. F.,** Nematophagous fungi with particular reference to their ecology, *Biol. Rev.,* 62, 245, 1987.

Chapter 2

FUNGI COLONIZING CYSTS AND EGGS

G. Morgan-Jones and R. Rodríguez-Kábana

TABLE OF CONTENTS

I. INTRODUCTION

Interactions between fungal antagonists of diverse biological capacities and nematodes in agricultural soils have long been known to exist. Fungi are thought to play a significant role in regulating nematode population dynamics and thus induce suppressiveness.

Obligate plant-parasitic nematodes belonging to the family Heteroderidae Skarbilovich are liable to be deleteriously affected by fungi in various ways at sedentary stages of their life cycles either within root tissue or, more often, when exposed on the root surface, or when free in the soil. In the genera *Globodera* Mulvey and Stone and *Heterodera* Schmidt, as females swell, break through the root cortex, molt, and become converted into cysts, they are increasingly vulnerable to invasion by opportunistic soil fungi present in the rhizosphere where fungal growth is stimulated and enhanced by root exudates. While the elongated neck region remains embedded and attached to root tissue, the cyst assumes an inflated form and becomes almost totally exposed, being more or less globose to pear-shaped in *Globodera* and lemon-shaped in *Heterodera*. In the former, the vulva and anus are located on the rounded cyst posterior while in the latter the vulva forms a terminal cleft and the anus is located anteriorly to it on the dorsal surface. In both cases these natural orifices are exposed to the soil mycoflora and provide a potential avenue for fungal invasion and eventual colonization. Following transformation of the female cuticle into a tough brown cyst and detachment from the root as a result of disintegration of root tissue, this reproductive structure, containing eggs, may remain in soil for an extended period before larval hatching and is vulnerable to invasion and degradation by fungi over time. Although cyst walls decompose comparatively slowly and thus provide good protection for the eggs contained within them, viability potential decreases the longer the eggs remain unhatched. The nematode cyst, in effect, provides a favorable ecological niche for fungi since the products derived from internal organ decomposition following egg differentiation offer a highly suitable nutrient supply for growth. As Thorne[1] has pointed out, cysts containing dead eggs and larvae are frequently filled with fungal mycelium.

In the genus *Meloidogyne* Goeldi, egg masses extruded in a gelatinous matrix surrounding the protruding posterior end of the female are susceptible to attack by fungi in much the same way as are the cysts, and eggs within them, in *Globodera* and *Heterodera*. Exposed in the host plant rhizosphere, eggs of *Meloidogyne* are, at the time of their extrusion, proximal to fungi growing in the vicinity of the root surface. If they remain unhatched for a time the likelihood of destruction by fungi is increased.

II. HISTORICAL REVIEW

That soil fungi colonize the reproductive structures, especially cysts and eggs, of plant-parasitic nematodes has been known for many years. As long ago as 1877, Kühn[2] reported discovering a fungal pathogen of females of the beet cyst nematode, *Heterodera schachtii* Schmidt, in Germany. The fungus was named *Tarychium auxiliarum* Kühn [later corrected by Kühn[3] to *Tarichium auxiliare*; now *Catenaria auxiliaris* (Kühn) Tribe]. In 1890, Hollrung,[4] in a general discussion of fungi in relation to sugar beet pests, mentioned three insect pathogens, *Isaria destructor* Metschn. [now *Metarhizium anisopliae* (Metschn.) Sorok.], and two species of *Entomophthora* Fres., in addition to the fungus encountered by Kühn. Baunacke[5] subsequently intimated that these infect *Heterodera* females, although, as Dollfus[6] has pointed out, he gave no references or account of an investigation substantiating this claim.

The first examination of cysts (of *H. schachtii*) was reported in 1929 by Korab[7] in the Ukraine. A fungus, named *Torula heteroderae* sp. nov., but not described [now considered to have possibly been *Phialophora malorum* (Kidd and Beaumont) McColloch], was said to cause browning, enlargement, and granularization of eggs and larvae. Cysts were found to bear brown necrotic patches and larval cuticle appeared nodulose due to the accumulation of spores on the surface. *Tarichium auxiliare* was found repeatedly in empty cysts following larval hatching and was thought to grow on residual slime. Another fungus, designated *Protomyces* sp. [now thought to have probably been a species of *Endogone* Link], was also found often in empty cysts. In addition, Korab noted infrequent presence of a number of globose, hyaline, thick-walled structures, referred to as spores, within eggs, and proposed the name *Olpidium nematodae* sp. nov. for them in spite of the fact that no zoospores were known to be produced. Tribe[8,9] has subsequently questioned the application of the generic name *Olpidium* (Braun) Rabenh. to these and suggested that the fungus may have belonged to the genus *Pythium* Pringsheim. *Arthrobotrys oligospora* Fres., a predacious species, together with a species of *Isaria* Pers. [a genus containing insect pathogens now classified in part as section *Isarioidea* Samson of the genus *Paecilomyces* Bainier], and a *Pythium* were encountered occasionally. Although over a quarter of a million cysts were examined during the study, no estimation of the overall level of fungal occurrence was given, but in some samples over 90% of the cysts were found to be colonized by the so-called *Torula*.

Some 3 years after the publication of Korab's findings, Goffart[10] presented the results of an investigation of cysts of the cereal cyst nematode, *Heterodera avenae* Wollenw., in Germany. Considerably higher levels of cyst colonization were reported to occur in this nematode than in beet and potato cyst nematodes. Up to 25% of all cysts extracted from the 0 to 20 cm layer of soil contained fungal mycelium. Fungus-occupied eggs were described as being frequently black in color and opaque in appearance, with some embryos destroyed. Eggs were also seen engulfed by superficial fungal hyphae covering the egg shells, in which case hatching was prevented from occurring. *Cylindrocarpon radicicola* Wollenw. [now *Cylindrocarpon destructans* (Zinssm.) Scholten, the anamorph of *Nectria radicicola* Gerlach and Nilsson] was isolated and assumed to be the fungus involved.

Rademacher and Schmidt[11] examined cysts of *Heterodera schachtii*, also in Germany, and reported that a much higher number of diseased cysts (43% as opposed to 15%) were found in samples from soils which contained few cysts as compared to soils with many cysts. *Metarhizium anisopliae* [as *Isaria destructor*] was determined to be the main fungus implicated in disease induction, but it was not isolated.

A further investigation of cysts of *H. schachtii* was conducted by Rozsypal[12] in Moravia. Between 51% and 64% of all cysts examined were reported to be diseased. Three fungi were encountered, with one predominating. Eggs colonized by the predominant fungus appeared black and its hyphae were said to form rows of globose, brown, thick-walled cells. It was, according to Schol-Schwarz[13], first identified by Jaczewski in a letter to Rozsypal, as *Torula heterodera* Korab, but van Beyma subsequently reidentified it for him as *Trichosporium populneum* Lambotte and Fautrey and this is the name used for it in his paper. Van Beyma[14] later realized that it was not *T. populneum* and transferred it to *Cadophora* Langer. and Melin [a synonym of *Phialophora* Medlar] as *Cadophora heteroderae* (Jacz.) Beyma [an incorrect author citation since the basionym was *Torula heteroderae* Korab.]. Later still van Beyma[15] proposed the new combination *Phialophora heteroderae* (Korab) Beyma for it. Schol-Schwarz[13] considered it to be *Phialophora malorum*; that is, the same fungus that Korab probably encountered earlier. Tribe,[9] however, has expressed doubt about the true identity of Rozsypal's fungus. The two additional fungi, found rarely by Rozsypal, were identified as *Protomycopsis* sp. and *Olpidium nematodae* Skvortzow, respectively. Tribe,[9] on the basis of the description and illustrations provided by Rozsypal, suggested the former to be a mycorrhizal *Endogone* [as *Glomus*]. The latter, represented by resting spores occupying a whole cyst rather than eggs, was considered by Tribe[8] to differ from the fungus given the same binomial by Korab (see above) and to possibly represent a species of *Entomophthora*.

Jones[16] made incidental reference to a disease of *H. schachtii* cysts in England but no fungi were isolated or identified. In some samples up to 50% of cysts examined were found to be affected. The most common causative agent, said to be probably a fungus, turned eggs black, and numerous dark-walled spores were observed filling them. The identity of two other fungi present, found only in single cysts, has subsequently been considered. One was probably an *Endogone*, since it was said to resemble the microcysts reported by Triffit.[17] These, which were often confused by nematologists with cysts of *Heterodera* and occurred commonly in various English soils, were thought by Gerdemann and Nicolson[18] to have been spores of that genus. Tribe[9] realized that the ornamented spores described belonged to *Tarichium auxiliare*.

Cysts of the potato cyst nematode, *Globodera rostochiensis* (Wollenw.) Stone [as *Heterodera rostochiensis* Wollenw.] were examined in the Netherlands by van der Laan[19,20] in the 1950's. His samples came from various countries and a number of fungi were found associated with them. *Phoma exigua* Desm. [as *Phoma tuberosa* Melhus, Rosenbaum and Schultz] was reported in the first study and this was followed by records of six other fungi. *Colletotrichum coccodes* (Wallroth) Hughes [as *Colletotrichum atramentarium* (Berk. and Br.) Tabenhaus] was found to occur as sclerotia on the cyst walls in material from the Netherlands, and *Humicola grisea* Traaen [as *Monotospora daleae* Mason] was isolated consistently. The latter was also encountered in material originating in Denmark. The presence of the *Colletotrichum* in the cysts was typified by small, black spots. *Pseudoeurotium ovale* Stolk and *Penicillium dangeardii* Pitt [as *Penicillium vermiculatum* Dang.] were reported from English samples, and *Xanthothecium peruvianum* (Hansen) Arx and Samson [as *Anixiopsis stercoraria* Hansen], *Exophiala jeanselmei* (Langer.) McGinnis and Padhye var. *heteromorpha* (Nannf.) de Hoog [as *Margarinomyces heteromorpha* Mangenot], and a species of *Scopulariopsis* Bainier, were found in samples from Peru. In the survey work, whole, surface-sterilized cysts were plated into agar. Because of this no assessment could be made of the activity of the individual taxa within cysts. Whether egg penetration and colonization occurred or the fungi grew as saprophytes on mucilage surrounding the eggs was not determined.

Populations of cysts of *Globodera rostochiensis* [as *Heterodera rostochiensis*], *H. schachtii*, and *H. avenae* Wollenw., were examined for disease and for the presence of naturally occurring antagonists by Willcox and Tribe.[21] *Globodera rostochiensis* cysts from 20 different sources (19 from potato growing areas of England and 1 from Bolivia) revealed very low levels of disease. No cysts containing an appreciable number of dead eggs were encountered. This corroborated unpublished observations made previously in England by F. G. W. Jones, which indicated little disease in *G. rostochiensis* cysts. A small sample of *H. avenae* cysts also showed a relatively low level of disease although some were found to be extensively damaged and colonized by an *Endogone*. Cysts of *H. schachtii* had higher disease incidence, and a diversity of unnamed fungi were said to have been observed associated with them. Cysts with a majority of dead eggs mostly yielded *Verticillium chlamydosporium* Goddard [as *Cephalosporium* sp.]. This species was isolated from both eggs and larvae. The only other fungus isolated with any frequency was an undetermined species of *Penicillium* Link which was found in some cysts containing appreciable numbers of diseased eggs.

Burnsall and Tribe[22] examined cysts of *H. schachtii* from an English soil and found just over 20% to be partially or substantially diseased. Eggs and larvae were colonized by four principal fungi, namely, *Cylindrocarpon destructans* (Zinssm.) Scholten, *Verticillium chlamydosporium*, a dark, sterile, crystal-forming fungus, and a fungus said to be similar to *Phialophora malorum* and *Exophiala mansonii* (Castellani) de Hoog [as *Rhinocladiella mansonii* (Castellani) Schol-Schwarz] and referred to as a ''black yeast''. Eggs infected by *C. destructans* were reported to be reddish brown in color due to the content of resting hyphae, and those colonized by the ''black yeast'' were partly or wholly filled with dark hyphae. Fifteen additional fungi were also recovered from eggs, the most common of which were *Fusarium tabacinum* (van Beyma) W. Gams and two species of *Phoma* Sacc. The remainder were not named. Some eggs occasionally contained large, yellowish, reticulately ornamented spores, similar to those of *Tarichium auxiliare*, but larger.

In the early 1970s, during the course of field trials on *H. avenae* in southeastern England, Graham and Stone[23] found nearly 40% of new cysts examined from one site to be diseased and mostly colonized by unidentified fungi. Chlamydospores of *Verticillium chlamydosporium* were recognized in some, and granular and shriveled eggs resembling those of *G. rostochiensis* infested by *Phialophora malorum*, as illustrated by van der Laan,[20] were seen.

Kerry[24] attempted to determine if fungi were involved in the known phenomenon of *H. avenae* decline in cereal monoculture in England. Infested soil was collected in the spring and sown with barley in the greenhouse. *Heterodera avenae* cysts were harvested in the summer and examined for the presence of fungi. Four species were encountered associated with disease of females and eggs. An *Entomophthora*-like fungus [later named *Nematophthora gynophila* Kerry and Crump] was the most predominant fungus killing females on roots, and this was also found infecting eggs within cysts. *Verticillium chlamydosporium* was commonly found in new cysts, often killing all of the eggs. *Tarichium auxiliare* and *Cylindrocarpon destructans* were also recovered, but few females or eggs were affected by them. In further studies on *H. avenae* and other cyst nematodes, Kerry and Crump[25] confirmed the involvement of these four fungi in female and egg pathology. *Verticillium chlamydosporium* was again the most frequently observed egg parasite, killing 50% of the eggs in females on barley roots in a field trial. The *Entomophthora*-like fungus was shown to attack a number of species of *Heterodera* but not *G. rostochiensis*.

In 1978, Goswami and Rumpenhorst[26] reported an unidentified fungus, together with *Fusarium oxysporum* Schlecht. and *F. solani* (Mart.) Sacc., associated with cysts of *G. rostochiensis* in West Germany. Judging from the available description and its morphology within eggs as seen in the published illustrations, the unidentified fungus, which was not isolated in pure culture, probably belonged to the *Exophiala*-complex. A similar level of colonization to that found by van der Laan[20] was reported. The *Fusarium* isolates could not, in inoculation experiments, be shown to be capable of parasitizing eggs, and it was concluded that the unnamed fungus was largely responsible for destruction of larvae within eggs. The fungus was reported to infect various pathotypes of both *G. rostochiensis* and *G. pallida*.

Additional accounts of the extent of disease and disorders occurring in cysts and eggs of *H. schachtii* and *H. avenae* were presented by Tribe.[9,27] A large number of samples were obtained from diverse geographical sources, including England, continental Europe, and the U.S. The level of disease of *H. schachtii* cysts from England, Germany, Italy, and the Netherlands was about 15%, although in one long term sugar-beet monoculture trial it was twice that level. None of the cysts of this nematode of American origin had any disease attributable to fungi, although some appeared to be physiologically disordered. A number of fungi were reported to be implicated, to various extents, in disease induction. *Verticillium chlamydosporium* and an entity referred to as "contortion fungus" were listed as the principal egg pathogens present, being widespread in distribution and frequent in occurrence within the cysts examined. The former was described as growing vigorously in mucilage within young cysts, forming a mass of hyphae in colonized eggs and invading unhatched larvae inside egg shells. The latter was designated a "contortion fungus" because of a characteristic contorted appearance assumed by larvae invaded by it. It could not be identified due to lack of sporulation following isolation in vitro and no reproductive structure was encountered in vivo. The fungus was reported to be extremely slow growing. Colonization of cysts by it was said to likely lead to blockage of cyst fenestrae by mycelium and its slow growth might allow time for some larvae to hatch. Such larvae could not escape from the cyst and would be particularly vulnerable to invasion by hyphae. Partly hatched larvae invaded by the fungus were typically found. It was suggested that death of characteristically contorted larvae might result from the biosynthesis of a fungal toxin. Several other fungi were found to be fairly widespread, but occurred in only a comparatively small number of cysts. These were referred to as minor egg pathogens and included *Cylindrocarpon destructans*, *Exophiala pisciphila* McGinnis and Ajello, possibly [according to the author] *Exophiala mansonii*, and the crystal-forming fungus previously reported by Burnsall and Tribe.[22]

During the past decade considerable progress has been made in documenting the consistent association of a number of opportunistic soil fungi with cysts and eggs of cyst and root-knot nematodes. In California, Nigh et al.[28] found strains of *Fusarium oxysporum* and *Acremonium strictum* W. Gams to occur regularly in eggs from young cysts of *H. schachtii*. These fungi were thought to be responsible for some control of this nematode under certain field conditions. Species of *Alternaria* Nees, *Cephalosporium* Corda [= *Acremonium* Link, pro parte], *Chaetomium* Kunze, *Cylindrocarpon* Wollenw., *Fusarium* Link, *Penicillium*, and *Phoma* were recorded with low frequency in the same study.

Morgan-Jones and Rodríguez-Kábana,[29] in a preliminary study of fungi associated with cysts of *Heterodera glycines* from an Alabama soybean field soil, recorded the presence of nine species, of which *Exophiala pisciphila*, *Fusarium oxysporum*, and *F. solani* were thought to be implicated as egg pathogens. In addition to these, *Neocosmospora vasinfecta* E. F. Sm., *Verticillium leptobactrum* W. Gams, and species of *Alternaria*, *Penicillium*, *Phoma*, and *Trichoderma* Pers., were found.

In a more extensive study of fungal colonization of *H. glycines* cysts in Arkansas, Florida, Mississippi, and Missouri soils Morgan-Jones et al.[30] encountered a taxonomically diverse mycoflora. Genera commonly found in cysts from all localities included *Fusarium*, *Gliocladium* Corda, *Neocosmospora*, and *Phoma*. A species of *Stagonospora* Sacc., believed to be consistently invasive of *H. glycines* eggs, was also present in samples from each of the four states. Other fungi recovered with frequency from cysts in some, but not all soils surveyed, included species of *Chaetomium*, *Codinaea heteroderae* Morgan-Jones, *Exophiala pisciphila*, and species of *Thielavia* Zopf. Observations of individual cyst contents following dissection indicated penetration of eggs regularly by several fungi, including *C. heteroderae*, *E. pisciphila*, and *Verticillium lamellicola* (F. E. V. Smith) W. Gams.

Gintis et al.[31] examined young, cream-colored cysts of *H. glycines* from roots of soybean [*Glycine max* Merr.] planted in infested soil samples from four North Carolina counties to attempt to determine whether or not cysts are colonized soon after they become exposed to the soil in the rhizosphere. Previous studies had concerned older, brown, detached cysts that had existed in soil for some time. Sixteen fungi were found and the level of fungal invasion varied with the soil sample. Overall, more than two-thirds of the cysts examined were found to bear fungal hyphae and signs of disease was prevalent in some. Five fungi, namely *Exophiala pisciphila*, *Fusarium oxysporum*, *F. solani*, *Neocosmospora vasinfecta*, and *Phoma leveillei* Boerema and Bollen occurred in all samples, the latter being the most abundant. Four other species of *Phoma* occurred in some of the samples.

A study by Stirling and Kerry[32] of brown cysts of *Heterodera avenae* collected during the summer in South Australia revealed the presence of *Verticillium chlamydosporium* [as *Diheterospora chlamydosporia* (Goddard) Barron and Onions]. Kerry et al.[33] further considered the role of soil fungi in natural control of *H. avenae* in England. A number of fungi were isolated from encysted eggs from three sites. *Verticillium chlamydosporium* was found to be the main parasite at all sites and this was thought to be in large part responsible for decline in nematode numbers. The only other fungus found to be present in more than 10% of the eggs was *Microdochium bolleyi* (Sprague) de Hoog and Hermanides-Nijhof. *Cylindrocarpon destructans*, *Gliocladium roseum* Bain, *Fusarium*, *Paecilomyces*, *Phialophora*, and *Phoma* spp., fungi known previously, from surveys by other researchers, to be frequently associated with cyst nematode reproductive structures, were recorded in lower numbers.

Gintis et al.[34] investigated the degree of colonization of *H. glycines* at various stages of development in and on soybean plant roots. Plants were grown in Alabama soybean field soil known to carry populations of the nematode. The study revealed increased colonization with time, both in terms of percentage of females and cysts invaded and in diversity of fungal species involved. Although some fungi occurred at a number of stages of cyst development, differences in the composition of the mycoflora were encountered from one stage to another. Examination of unexposed, sausage-shaped females within root tissue revealed few to be in any way affected by fungi. About 2% were found to be invaded, mainly by *Scytalidium fulvum* Morgan-Jones and Gintis and *Trichosporon beigelii* (Küchenm. and Rabenh.) Vuill. The low incidence of fungi indicated that the roots provided effective protection for the nematode. White, semi-inflated, lens-shaped, newly exposed females partly immersed in the roots were more readily colonized. About 20% of these were found to bear fungi. Half of fully-swollen, cream-colored cysts still attached to root surfaces were found to contain fungal hyphae. Fungi most frequently isolated from these stages of development were: *Chaetomium cochlioides* Palliser, *Exophiala pisciphila*, *Fusarium oxysporum*, *F. solani*, *Phytophthora cinnamomi* Rands, and *Trichosporon beigelii*. Older, detached, brown-colored cysts were found to be much more heavily colonized, with 70% containing fungal mycelium. *Fusarium oxysporum*, *F. solani*, *Neocosmospora vasinfecta*, *Phoma terrestris* Mont., and *Scytalidium fulvum* were the most abundant fungi present in these. Other species found with some frequency included *Paecilomyces lilacinus* (Thom) Samson, *P. variotii* Bain., and *T. biegelii*. Occurring at lower frequencies were *Cylindrocarpon tonkinense* Bugn., *E. pisciphila*, *Verticillium chlamydosporium*, and *V. lecanii* (Zimm.) Viegas. In addition, over 30 other fungi were found, but in low numbers. Cysts invaded by a number of the above mentioned fungi revealed eggs containing endogenous fungal hyphae. Eggs infected by *E. pisciphila* were filled with ellipsoid, pale brown, hyphal elements. Cysts from which *S. fulvum* was recovered contained torulose, brown hyphae and chlamydospore-like arthroconidia. *Paecilomyces lilacinus*, *P. variotii*, *V. chlamydosporium*, and *V. lecanii* were all found to deleteriously affect eggs. Eggs colonized by *P. variotii* were often swollen and discolored, some appeared lysed, while others contained large lipid globules. The contents of eggs invaded by *P. lilacinus* and *Verticillium* spp. were often consumed and replaced by fungal hyphae.

Dunn et al.[35] evaluated four isolates of *P. lilacinus* for their ability to colonize eggs of *Meloidogyne incognita* (Kofoid and White) Chitwood in vitro. Three of the isolates were found to colonize eggs by hyphal penetration. Dunn[36] reported a new species of *Paecilomyces*, *P. nostocoides* Dunn, from cysts of *Heterodera zeae* Koshy, Swarup and Seth, in Maryland, and the fungus was shown to be capable of penetrating egg shells of the nematode. Godoy et al.[37] considered it to be an aberrant form of *P. lilacinus* since it is identical to that species except for the presence of occasional, swollen, abnormal conidia in otherwise typical chains.

Morgan-Jones et al.,[38] through a survey of fungi associated with *H. glycines* cysts in the Cauca Valley, Colombia, attempted to determine if a newly-introduced nematode could be colonized by soil fungi indigenous to a given area. Tovar and Medina[39] had, the previous year, reported occurrence of *H. glycines* on soybeans in this locality for the first time. It was formerly unknown in South America. Previous reports, documented above, had been based on cysts derived from soils where particular nematodes and indigenous fungi had co-existed for a relatively long period of time. Nineteen different fungal species were encountered. *Fusarium oxysporum* and *F. solani* predominated, but *Geotrichum candidum* Link, *Gliocladium roseum*, *G. catenulatum* Gilman and Abbott, and *Paecilomyces lilacinus* occurred in significant numbers. The unnamed species of *Stagonospora* frequently found by Morgan-Jones et al.[30] in the U.S., was also present in Colombia. These results indicated that a nematode novel to an area is indeed vulnerable to colonization by fungi pre-existing in that environment.

Fungal parasites of *H. avenae* were isolated from four sites in southern Sweden by Dackman and Nordbring-Hertz.[40] In all, fifteen fungi were found in different stages of cyst development. *Verticillium chlamydosporium* was the most common species found in young cysts and a different, unnamed species of this genus was the most frequently occurring species in cysts from soil. A species of *Cylindrocarpon*, *Microdochium bolleyi*, and *Paecilomyces lilacinus* were also isolated from eggs derived from cysts of soil origin. In one sample, 43% of invaded eggs were found to contain *P. lilacinus*.

The considerable number of investigations conducted on the association between opportunistic soil fungi and cysts and eggs of plant-parasitic nematodes, reported above, concentrated predominantly on species of *Heterodera*. Comparatively little study in recent years had been made of cysts of species of *Globodera* until Morgan-Jones and Rodríguez-Kábana[41] examined populations of *G. pallida* (Stone) Behrens and *G. rostochiensis* in Peru, where the nematodes are native. In most previously published reports of fungi occurring in cysts of *Globodera*, low levels of colonization were found, for which no convincing explanation could be advanced.[20] An appreciable mycoflora, comprising 28 taxa was, however, found associated with cysts in Peru. Close to 40% of those examined yielded fungal colonies. The most abundantly occurring species was *Cylindrocarpon destructans*. Two other species of this genus, namely *C. didymum* (Hartig) Wollenw. and *C. gracile* Bugn., were also present, but in lower numbers. Species of *Fusarium*, *Gliocladium roseum*, and *Ulocladium atrum* Preuss were other fungi found with some frequency. *Paecilomyces lilacinus* and several other fungi reported previously from nematode cysts were rarely present.

Reports of fungi associated with eggs of *Meloidogyne* have also been relatively limited. As far back as 1938, Linford and Oliveira[42] listed over 50 enemies of *Meloidogyne*, including a single egg parasite, a species of *Penicillium*. Stirling and Mankau[43,44] described *Dactylella oviparasitica* Stirling and Mankau as a parasite of an unnamed species of *Meloidogyne* in California. In Peru, eggs of *M. incognita* var. *acrita* Chitwood, from potato roots, were found by Jatala et al.[45] to be heavily infected by *Paecilomyces lilacinus*. Godoy et al.[37] encountered *Fusarium oxysporum*, *P. lilacinus*, *Pseudopapulospora kendrickii* Sharma, and *Verticillium chlamydosporium* as egg parasites of *M. arenaria* (Neal) Chitwood in Alabama peanut field soil. *Paecilomyces lilacinus* occurred with the highest frequency, being present in 47% of the colonized eggs examined, followed by *V. chlamydosporium*. Eggs of *M. incognita*, derived from galled soybean plants at a single location in Alabama, were examined for the presence of fungi by Morgan-Jones et al.[46] The level of colonization found was lower than that recorded for *M. arenaria* by Godoy et al.,[37] but the mycoflora, comprised of seven species, was more diversified. *Aureobasidium pullulans* (De Bary) Arnaud, *F. oxysporum*, *Gliocladium roseum*, and *P. lilacinus* were present in substantial numbers.

Very few reports exist of fungi parasitizing eggs of other nematodes. Lysek[47] studied parasitism of eggs of *Ascaris* L., an animal parasite, by placing them in soil samples. A species of *Cephalosporium* and *Fusarium oxysporum* var. *redolens* (Wollenw.) Gordon [as *F. redolens* Wollenw.] were reported to destroy the eggs. In a later report, Lysek[48] listed *F. oxysporum* var. *orthoceras* (App. and Wollenw.) Bilay, *F. oxysporum* var. *redolens*, *F. solani* f. sp. *radicicola* (Wollenw.) Snyder and Hansen [as *F. javanicum* Koord., var. *radicicola* Wollenw.], and *Paecilomyces marquandii* (Massee) Hughes as principal parasites. A further study by Lysek[49] reported the ability of *Acremonium bacillisporum* (Onions and Barron) W. Gams, *Helicoon farinosum* Linder, *Mortierella nana* Linnem., *Paecilomyces lilacinus*, *Verticillium chlamydosporium*, and *V. bulbillosum* W. Gams and Malla to perforate egg shells and invade eggs. Barron[50] considered *Rhopalomyces elegans* Corda to be the only known bona fide egg parasite.

This review of studies spanning over a hundred years in various geographical areas on several nematode hosts, both cyst and root-knot, indicates the participation of a range of opportunistic soil fungi in the colonization, and in some cases, destruction of reproductive structures.

III. HOST-PATHOGEN RELATIONSHIPS

A number of fungi appear to occur consistently within cysts and/or eggs of plant parasitic nematodes, irrespective of host identity or geographical location. Many of these are ubiquitous in agricultural soils and cosmopolitan in distribution. Such taxa are frequently present in appreciable numbers, although the dominating fungus present in any given situation or area varies. Others occur sporadically in low numbers and can, at best, be only considered incidental. The presence of a particular fungus within a cyst, or among egg masses, however, does not necessarily mean an ability on its part to parasitize. Some, if not most taxa encountered, especially those occurring rarely, undoubtedly grow on mucilage between eggs within cysts. Many of these opportunistic fungi do, nevertheless, appear to be capable of inducing physiological disorder of eggs, abortion of embryonic development, and, in some cases, physical disruption. The occurrence of a restricted mycoflora consisting of much the same fungi in association with cysts and eggs in most of the surveys conducted probably indicates that some measure of specialization is a prerequisite to successful and full utilization of the nematode reproductive structures as a food source.

There does not appear to be any consistent pattern reflecting host differences or preferences. Genera of fungi such as *Cylindrocarpon, Exophiala, Fusarium, Gliocladium, Paecilomyces, Phoma,* and *Verticillium* Nees appear to be able to colonize cysts and eggs of a range of plant-parasitic nematodes. A number of parameters, including plant host, cropping history, soil peculiarities, and climatic conditions, are believed to be operative in determining the preponderance of one fungus or another in any given situation. There is some indication, for example, that a colder climate favors the occurrence of *Verticillium chlamydosporium,* whereas *Paecilomyces lilacinus* appears to be present more abundantly in warmer regions (the temperature/growth relationships of these fungi in vitro confirms this assumption).

Some of the anomalies extant in the literature concerning levels of fungal colonization of cysts may, at least in part, be explained by procedures used in the research. Quantitative studies on the degree of fungal occurrence in cysts recovered from soil may be affected by the conditions to which cysts and/or soil are submitted following collection. Dackman and Nordbring-Hertz[40] reported that rate of recovery of some fungi is altered when soil is stored for different periods of time. With longer storage at 2°C the frequency of occurrence of *Verticillium chlamydosporium* appeared to decrease, whereas another *Verticillium,* unidentified to species, was favored by the low storage temperature. It was suggested that if this were extrapolated to the field *V. chlamydosporium* may be an early colonizer of young cysts during warm periods in Sweden, with the other *Verticillium* becoming more important as soil temperatures decrease. Since periods and conditions of storage prior to examination have varied over the years among different researchers, some results concerning levels of fungal colonization should be treated with caution.

The disparities in reports of fungal invasion of *Globodera rostochiensis* cysts raise several questions concerning host-pathogen relationships. Tribe[51] discussed the then-assumed rarity of parasitism of *G. rostochiensis* and offered possible explanations for this. Two possible limiting factors were mentioned. These were the absence of a vulval cone and the fact that the vulval aperture in the cyst wall is small, together with the presence of an additional layer of cuticle in the female of this nematode as compared to species of *Heterodera.* A reduced aperture and the extra layer of cuticle were thought to offer effective barriers to penetration.

The results of van der Laan,[20] Goswami and Rumpenhorst,[26] and Morgan-Jones and Rodríguez-Kábana,[41] indicate that cysts of *Globodera pallida* and *G. rostochiensis* are sometimes invaded by soil fungi to much the same degree as those of *Heterodera,* at least in some agricultural soils. It seems possible, however, that in certain geographical areas where *Globodera* is not native, fungi fail to colonize cysts consistently and in significant numbers, but it does not appear to always follow that where a nematode is newly introduced colonization levels will be low. The study by Morgan-Jones et al.[38] on *H. glycines* in Colombia indicates otherwise.

Evidence has accumulated that suggests varying capacities to colonize eggs within particular fungal species. Differing biotypes possessing divergent physiological capabilities in all likelihood exist and some of these are able to colonize particular ecological niches such as nematode cysts and eggs while others cannot. It seems reasonable to assume that differing selection pressures, including ones related to races of nematodes and resistance factors, favor the dominance of one or other biotype in any given microhabitat. Differences in biotypes may well partly account for the different degree of colonization of *Globodera* cysts in Peru and Europe. In some circumstances where a nematode is not native an extended time lapse may well precede the emergence and build-up of fungal biotypes adapted to the host.

Species of *Fusarium* provide an example of taxa for which no generalizations can be made regarding ability to parasitize nematode eggs. Members of this genus have frequently been encountered as colonizers of nematode cysts. Examination of eggs from cysts where *Fusarium* is present, however, often gives conflicting results as to the capacity of individual isolates to perforate the shells and destroy the egg contents.[41] Nigh et al.[28] convincingly documented the ability of isolates of *F. oxysporum* from *H. schachtii* to parasitize eggs of that nematode, and similar observations have been made on strains of both *F. oxysporum* and *F. solani* isolated from *H. glycines* by Gintis et al.[34] and Morgan-Jones and Rodríguez-Kábana.[41] In vitro studies by Goswami and Rumpenhorst[26] and Godoy et al.[52] however, failed to demonstrate the ability of *F. oxysporum* and *F. solani* isolates derived from cysts to parasitize eggs. In the latter study the two species were tested against both *H. glycines* and *Meloidogyne arenaria*. It should be noted that in the same experiment both *Verticillium lamellicola* and *V. leptobactrum* were found to parasitize eggs. Tribe[9] relegated *Fusarium* among miscellaneous fungi considered to be at most only weakly parasitic, although recognizing that species of the genus can be present endogenously within eggs. The interplay betwen nematode hosts and infraspecific biotypes of a genus such as *Fusarium* is likely to be complex and changing. The condition of eggs, particularly degree of maturation, when a given fungus makes contact, and other factors, will in part presumably dictate ensuing parasitic activity, if any.

Rodríguez-Kábana et al.[53] reported differences in virulence of individual isolates of *Paecilomyces lilacinus* against nematode eggs. Isolates of this species appear to differ widely in their ability to establish in soil, and thereby in their capacity to exercise a biocontrol effect on plant-parasitic nematodes. Genetic differences among strains or biotypes may be decisive in determining their relative ability to parasitize cysts and eggs. Similar variation is also known among biotypes of *Gliocladium roseum*.[53] Recent data obtained in our laboratory[77] indicate a correlation between the ability to biosynthesize a diffusible, possibly toxic, yellow pigment in agar cultures and biocontrol efficacy on the part of different isolates of this fungal species.

Overall interactions between opportunistic soil fungi and their nematode hosts appear to be quite variable. The main types of destructive activity are thought to be direct physical perforation of such entities as egg shells and larval cuticles, enzymatic dissolution of nematode structural elements, and physiological disturbances brought about by biosynthesis of diffusible toxic metabolites.[54-56] Fungi may also play a role in long term degradation of cyst exocuticle in soil.[30] In terms of activity, fungi can be broadly categorized into three groups, although individual species might overlap in their involvement. Firstly, fungi that are able to enter cysts early, grow saprophytically on cyst contents including mucilage surrounding eggs and, possibly, the multilayered endocuticle but leave unaffected the thick, resistant exocuticle and the egg shells enclosing the larvae. Secondly, fungi that are capable of perforating egg shells and larval cuticles within, and thereby arrest the reproductive process. Thirdly, a succession of fungi that are probably involved in the degradation of cysts in soil over time. Fungi belonging to any or all these categories can deleteriously affect nematode reproduction by chemical means. Additional effects might involve indirect modification of the environment through the presence of fungal biomass. For example, by the depletion of the oxygen supply available to juveniles. In any given situation, as alluded to briefly above in connection with *Fusarium*, fungal activity bringing about a detrimental effect will depend on several factors. These include the stage of nematode development when a fungus is in immediate proximity, the amount of fungal biomass, the ability to exogenously alter the nematode's condition by translocatable metabolites, either enzymatic or toxic, and the capacity for physical disruption.

The cyst, as a protective entity, is clearly only partly successful in shielding eggs since fungal hyphae can apparently enter through the natural openings, anal aperture and vulva, which remain from its life as a female. Once inside a cyst, fungal biomass increases due to readily available nutrients derived from decomposition of the internal organs of the female.

Whether within a cyst or in an extruded mass enveloped by a gelatinous sac, as in *Meloidogyne*, eggs are frequently exposed to, and are often in relatively close contact with, fungal hyphae. Although an egg is the most resistant stage in a nematode's life cycle to natural environmental stress, some fungi appear to be able to overcome the substantial protection afforded juveniles by the enveloping shell. The vulnerability of eggs to fungal colonization appears to vary considerably according to age, although in aggressive parasites such as *Paecilomyces lilacinus*, there are no apparent differences in level of invasion among various developmental stages.[55] It is possible that, in the case of weak parasites, unless the fungus reaches eggs at a very early stage of differentiation, no penetration of the shell is effected. In an immature condition, when the egg is filled with undifferentiated granular material and the shell layers are not fully elaborated, it is in its most vulnerable condition. A fungal hypha coming into contact with a very young egg is able to forcibly buckle the shell and to eventually penetrate by a relatively wide, irregular rupture.[54,56] As the egg matures and cleavage of its contents begins, hyphae cannot readily perforate the shell and do so only through a narrow pore-like opening.[55] In vitro experiments reported by Morgan-Jones and Rodríguez-Kábana[56] have indicated that young eggs are more readily colonized than older ones, particularly those containing a full-differentiated juvenile. It is not known if any enzymatic weakening of the shell precedes hyphal penetration, but this is more likely to be a requirement for entry into mature eggs than into young ones.

Morgan-Jones et al.[54] have conclusively demonstrated, by in vitro and ultrastructural studies, the ability of *Verticillium chlamydosporium* not only to prevent hatching but to colonize eggs and juveniles of *Meloidogyne arenaria* by hyphal penetration. In water agar plates inoculated with the fungus, up to 70% of introduced eggs were found to be parasitized after 10 days. Sections showed hyphae to be present within both infected eggs and enclosed juveniles. The presence of fungal hyphae in close proximity to the external surface of egg shells and within the venters resulted in visible structural changes. The outer, vitelline layer of the shell in part disintegrated, producing pore-like gaps, and some dissolution of both chitin and lipid layers became evident. Fungal hyphae entered juveniles by broad breaks in the cuticle. Juveniles occupied by hyphae became necrotic and no trace of somatic musculature could be seen. The cuticle of diseased larvae became folded, and individual cuticular layers and the hypodermis disappeared. In a similar study on *Paecilomyces lilacinus* Morgan-Jones et al.[55] showed that eggs of *M. arenaria* are readily parasitized in vitro by the fungus at various stages of development. These included some at a two to eight-cell embryonic stage, some intermediate in development, and some containing fully-formed juveniles. Invaded eggs became appreciably swollen, indicating changes in shell permeability. When a hyphal tip came into contact with an egg shell, two changes in configuration frequently occurred. The apical area of the hyphae became inflated and some buckling of the shell occurred peripheral to the area of contact, especially in young eggs. Eventually the hyphal tip became truncate and closely appressed to the shell surface, following which the vitelline layer was pierced by a narrow penetration peg. The entering hyphae assumed normal dimensions once beyond the vitelline layer resulting in the forcible breakdown of wide areas of the chitin and lipid layers. As a result of fungal occupation of eggs three structural changes in the shells became apparent. The vitelline layer split into three constituent bands, numerous vacuoles appeared in the chitin layer, and the lipid layer largely disappeared.

The ability of eggs to withstand adverse environmental stresses depends to a large degree on the impermeability of their shells. Bird and McClure[57] report permeability of the tylenchoid egg shell to be regulated by the lipid layer which lies internal to the vitelline and chitin layers. Shells of immature eggs at the granular stage may not be entirely impermeable. Should diffusible toxic fungal metabolites reach such eggs and inward seepage occur, some physiological disorder could be expected to follow. The same effect would result, irrespective of the stage of egg development, should shell permeability be altered by hydrolysis of key elements in its composition.

Many of the fungi found associated with cysts and eggs are known to be chitinolytic and it seems possible that enzymatic activity and inward seepage of toxic metabolites can be involved in inducing disease. Tribe[9,27] and Morgan-Jones and Rodríguez-Kábana[29] have reported many cysts of *Heterodera* spp. to frequently contain lysed, shriveled, coagulated, or decayed eggs. A condition referred to by Tribe[9] as "oily degeneration" is occasionally encountered. In such eggs, large oily globules are present, sometimes within healthy-appearing juveniles or, more often, where the contents have degenerated. In cysts containing eggs in these conditions, fungal mycelium is usually present. Frequently, however, no hyphae appear to be present within such eggs. This suggests that a detrimental exopathic effect is achieved without hyphal entry and physical disruption.

The presence of some fungi has been reported to have a stimulatory effect on larval hatching, while others have an inhibitory effect.[56] Hatching of plant-parasitic nematodes is known to occur in response to stimuli provided by exudations from roots of host plants. Nematodes such as *Globodera rostochiensis* respond to a narrow range of hatching factors[58] while others, such as *Heterodera schachtii*, are induced to hatch by a wide range of agents.[59] Hatching factors induce juveniles to secrete enzymes that weaken the shell and facilitate their exit. This enzymatic activity alters permeability, as does pre-emergence larval movement, which emulsifies the lipid layer. Inward passage of fungal metabolites can bring about a number of responses. If toxic, hatching is prevented, leading to the demise of juveniles. If nontoxic, but simulatory, juvenile emergence can be enhanced. All told, a number of biochemical interactions between fungi and nematode eggs probably occur.

IV. IMPLICATED PATHOGENS

As indicated above, a sizeable number of fungal species have been recorded as occurring in cysts of *Globodera* and *Heterodera* and a few have been found to parasitize eggs of these genera and those of *Meloidogyne*. The majority of fungi found within cysts cannot, by all indications, parasitize eggs and most have only rarely been encountered. A number have, however, been found consistently within cysts in significant numbers and some are known to penetrate egg shells. Comments on those considered to play a significant role in cyst and root-knot nematode pathology follow.

A. *Cylindrocarpon* spp.
Several species of this genus, including *C. destructans*, *C. didymum*, *C. gracile*, and *C. tonkinense*, have been recovered from cysts of the Heteroderidae.[22,33,34,41] In addition, un-

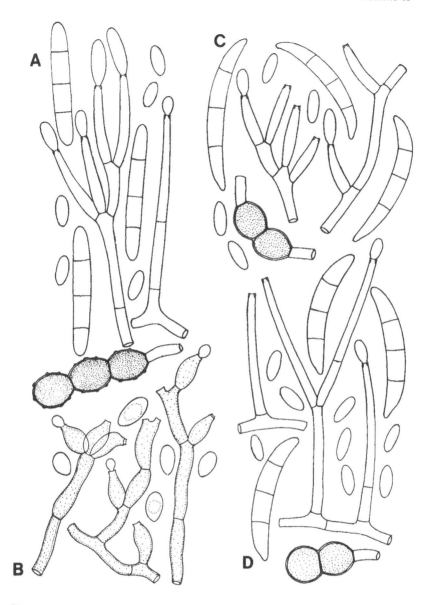

FIGURE 1.   (A) *Cylindrocarpon destructans*; (B) *Exophiala pisciphila*; (C) *Fusarium oxysporum*; (D) *Fusarium solani*.

determined species have been reported from eggs of *H. schachtii* in California[28] and *H. avenae* in Sweden.[40] *Cylindrocarpon destructans* (Figure 1A) is the species most commonly encountered. First reported as a parasite of *H. avenae* in Germany by Goffart[10] in 1932, it has subsequently been shown to be widespread in occurrence in cysts and was the fungus having the highest frequency in the study by Morgan-Jones and Rodríguez-Kábana[41] on *Globodera* in Peru. Booth[60] reported an isolate from eggs of *G. rostochiensis* in Scotland. It has also been found, at low frequency, in cysts of *H. schachtii* in England[22] where it was noted to be capable of penetrating eggs. Infected eggs were described as being reddish-brown in color due to the content of endogenous hyphae and it was reported to give a similar appearance in the Peru material.[41] Kerry[24] and Kerry and Crump[25] found it, again in low numbers, in females and eggs of *H. avenae* in England. Although it occurred in nematode-suppressive soils it was not considered important because of its scarcity. Tribe[27] noted *C.*

destructans to be reasonably constant in occurrence, though at low frequency. Van der Laan,[20] although he did not isolate *C. destructans* from cysts of *G. rostochiensis*, tested the ability of an isolate of it, of unspecified origin, to colonize cysts and penetrate eggs on agar medium. The results were negative however. Hams and Wilkin[61] used *C. destructans* in trials against both *G. rostochiensis* and *H. avenae* and their results were also negative. This fungus is predominantly found in soil colonizing senescing plant roots,[62] but some isolates are known to be pathogenic to plants.[60] Its role in cyst nematode pathology is uncertain, although there can be no doubt of its capacity, in some circumstances, to destroy eggs. It is possible that in particular instances, such as in potato fields in Peru infested by *G. pallida* and *G. rostochiensis*, a biotype of this species, capable of parasitizing eggs of the nematodes, has been selected for and predominates.[41] In cysts of *Globodera* colonized by *C. gracile*, pale brown chlamydospores were observed within eggs, but in the case of *C. didymum* no apparent hyphal penetration of eggs occurred.[41]

B. *Exophiala-Phialophora* complex.

A number of fungi, sometimes referred to as "black-yeasts", that produce brown or greenish-brown to black colonies and small, unicellular, ellipsoid to oblong conidia, have been reported from cysts and eggs by various researchers. Some confusion exists, however, as to their proper identity in older reports. As indicated in the above historical review, Korab[7] encountered what was possibly *Phialophora malorum* in cysts of *H. schachtii*, and Rozsypal[12] found a similar fungus in association with the same nematode. Although Schol-Schwarz[13] named the latter *P. malorum* also, Tribe[9] remained unconvinced of its true identity since Rozsypal's illustrations, which were more characteristic of an *Exophiala*, failed to recognize the distinctly collared conidiogenous cells typical of *P. malorum*. Van der Laan[20] found *Exophiala jeanselmi* var. *heteromorpha* [as *Margarinomyces heteromorpha*] in cysts of *G. rostochiensis* from Peru. Goswami and Rumpenhorst[26] and Tribe[27] also reported "black-yeasts" from cysts of *G. rostochiensis* and *H. schachtii*, respectively. The former authors considered their fungus to be similar to *P. malorum* while two of Tribe's isolates were identified as *Exophiala pisciphila* (Figure 1B). It may well be that the earlier researchers mistakenly reported *P. malorum* to be involved and that most, if not all, the records were actually based on species of *Exophiala* Carm. Morgan-Jones and Rodríguez-Kábana[29], Morgan-Jones et al.,[30] and Gintis et al.[31,34] reported *Exophiala pisciphila* to occur relatively consistently in cysts of *H. glycines* in the southeastern U.S. Its presence is indicated by inflated hyphal elements in the form of chains of pale to mid brown cells within invaded eggs.[29] First reported as the causal agent of systemic mycosis of catfish in Alabama,[63] it has also been isolated from a diseased swim bladder of red snapper.[29] Although commonly soil-borne,[64] and able to parasitize eggs, it generally occurs in low numbers in cyst samples. This probably reflects an inability to effectively compete in the soil ecosystem. Morgan-Jones et al.[30] reported other "black yeasts", including possibly *E. jeanselmei* [lack of conidiation precluded identification of some isolates], within cysts of *H. glycines*. Tribe[9] expressed the view that species of *Exophiala* other than *E. pisciphila*, including possibly *E. mansonii*, which was isolated regularly from wheat field soils in Germany by Domsch and Gams,[65] may also be involved in egg pathology.

C. *Fusarium* spp.

A number of species of *Fusarium* have been recorded as occurring in cysts, including *F. equisetii* (Corda) Sacc., *F. lateritium* Nees, *F. moniliforme* Sheld., *F. oxysporum*, *F. semitectum* Berk. and Rav., *F. solani*, and *F. tabacinum*.[22,26,28,30,31,34,41] Of these *F. oxysporum* and *F. solani* are by far the most frequently encountered. Although these fungi regularly enter and colonize cysts, there is, as alluded to above, no evidence that they are always able to effectively and consistently parasitize eggs. While some biotypes of *F. oxysporum*

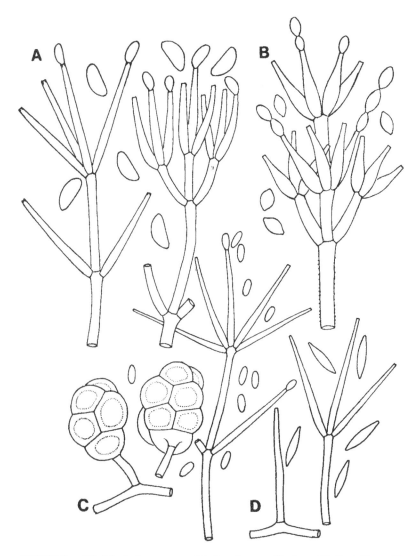

FIGURE 2.   (A) *Gliocladium roseum*; (B) *Paecilomyces lilacinus*; (C) *Verticillium chlamydosporium*; (D) *Verticillium lamellicola*.

(Figure 1C) and *F. solani* (Figure 1D) appear to be capable of penetrating eggs, in most instances eggs within cysts colonized by *Fusarium* spp. remain unaffected. Occasionally, eggs in such cysts appear abnormal, some being inflated (indicating a change in shell permeability), lysed, or shriveled, while others enclose larvae containing large, irregularly-shaped oil drops.[30] The presence of *Fusarium* hyphae within cysts, particularly those containing very young eggs, may physiologically disorder embryonic development through enzymatic and/or toxic effects. *Fusarium* spp. are known to produce a range of toxic metabolites, including trichothecenes.[66]

D. *Gliocladium* spp.

Both *Gliocladium catenulatum* and *G. roseum* have been recorded to occur in cysts of *H. avenae* and *H. glycines* in Europe and North America,[30,33,34] and in cysts of *Globodera* in Peru,[41] sometimes in appreciable numbers. As in the case of species of *Fusarium*, some strains appear to deleteriously affect eggs, while others are ineffective in this regard. *Gliocladium roseum* (Figure 2A) is a common soil fungus and is known to be a destructive mycoparasite.

E. *Paecilomyces* spp.

Paecilomyces lilacinus and *P. variotii* have been reported to occur in cysts of *Globodera* and *Heterodera* in various geographical locations,[34,35,40,41] and the former is known from eggs of *Meloidogyne arenaria* and *M. incognita* in North America and Peru.[37,45,46] *Paecilomyces lilacinus* (Figure 2B), although known to occur on insects in the tropics, is typically soil-borne and is considered to be an important egg pathogen,[67] at least in some instances. It is common in the rhizosphere of a number of plants,[67] and produces the antibiotics leucinostatin and lilacin.[68] Murao et al.[69] have reported biosynthesis by *P. lilacinus* of a cell wall lytic enzyme, capable of lysing the chitin-containing walls of *Rhodotorula* and *Sporobolomyces*. This enzyme may be involved in enabling *P. lilacinus* to colonize nematode eggs whose shells contain chitin.[59] This fungus has been used in biocontrol of the potato tuber moth because of its ability to parasitize the larval stage,[70] and it is known to cause several mycoses in man.[71,72]

F. *Verticillium* spp.

Six species of this genus, all belonging to section *Prostrata* W. Gams, namely *V. catenulatum* (Kamyschko) W. Gams, *V. chlamydosporium*, *V. lamellicola*, *V. lecanii*, *V. leptobactrum*, and *V. psalliotae* Treschow have been recorded as occurring in cysts and/or eggs of *Globodera*, *Heterodera* and *Meloidogyne*.[9,22,24,29,32-34,37] In addition, a seventh member of this section, *V. bulbilosum*, has been shown to be capable of penetrating egg shells of *Ascaris*.[49] *Verticillium psalliotae* has been isolated from cysts of *Heterodera mothi* Khan and Husain from Georgia.[79] Bursnall and Tribe[22] considered *V. chlamydosporium* (Figure 2C) to be the principal egg parasite of *H. schachtii* in England, and Kerry[24] thought it to be possibly involved in inducing decrease of *H. avenae* populations in cereal monoculture. Godoy et al.[52] tested in vitro the capacity of 14 taxonomically diverse fungal species isolated from cysts of *H. glycines* to invade eggs of that nematode and those of *Meloidogyne arenaria*. *Verticillium lamellicola* (Figure 2D) and *V. leptobactrum* were found to colonize between 60 and 98% of the eggs to which they were exposed and were the most effective in this regard among the fungi included in the experiments. It seems likely that many of the species of *Verticillium* classified in section *Prostrata* play a significant role as parasites of cysts and eggs.

V. BIOCONTROL POTENTIALS

The feasibility of using a number of the opportunistic soil fungi discussed above for biocontrol of plant-parasitic nematodes is currently being explored. Preliminary studies have indicated that *Gliocladium roseum*, *Paecilomyces lilacinus*, and *Verticillium chlamydosporium* have some potential for controlling *Heterodera avenae* and *Meloidogyne* spp. under greenhouse and/or field conditions.[25,37,73,74] Jatala et al.,[73] found that potato plants grown in plots heavily infested with *M. incognita* and inoculated with *P. lilacinus* had a significantly lower root galling index than those grown in plots to which organic matter and nematicides were applied. Over 80% of egg masses derived from plants grown in *P. lilacinus* treated soil were found to be infected by the fungus and over 50% of the eggs were destroyed. In experiments to determine effects of multiple *P. lilacinus* application Jatala et al.[74] found that one application was sufficient to establish the fungus. After three crops the galling index remained substantially reduced. Dickson and Mitchell[75] reported *P. lilacinus* to be largely ineffective for the management of *M. javanica* (Treub) Chitwood on tobacco in microplots. Reduction in the number of galls caused by *M. arenaria* on roots of summer squash [*Cucurbita pepo* L., var. *melopepo* (L.) Alef.] in greenhouse tests following addition of oats colonized by two isolates of *Gliocladium roseum* and one each of *G. catenulatum*, *P. lilacinus*, *V. chlamydosporium*, and *V. lamellicola* were reported by Rodríguez-Kábana et al.[53] Two other isolates of *P. lilacinus* were ineffective in reducing root galling. Species of *Paecilomyces* and *Verticillium* were not recovered from females obtained in galls of plants from soil treated with members of these genera, whereas *Gliocladium* species were found to be present in females from soil to which they had been added. In contrast to *P. lilacinus*, isolates of *Gliocladium* performed well in controlling *M. arenaria* and in their ability to colonize soil. They were consequently considered to have the most promise as biocontrol agents. Culbreath et al.[76] studied the effect of *P. lilacinus* treatments, with and without chitin amendments, on degree of galling by *M. arenaria* in squash followed by tomato [*Lycopersicon esculentum* Mill.]. Numbers of galls in squash after 6 weeks were not affected. Tomatoes planted in the same soil following squash were found, however, after a further 6 weeks, to have less galling and fewer juveniles/gram of roots in chitin/*P. lilacinus* treatments than in controls. Results indicated that combinations of chitin and *P. lilacinus* amendments were effective in controlling *M. arenaria*, but the authors did not interpret the decline in the nematode populations after tomato as due solely to the activities of *P. lilacinus*. It was thought that a mycoflora consisting of a number of organisms antagonistic to the nematode probably developed in response to the use of chitin amendments and that these, at least in part, contributed to the induced suppressiveness. Kerry et al.[77] contributed observations on the introduction of *V. chlamydosporium* into soil for control of *H. avenae*. Some isolates of this fungus added to soil on ground oat grain reduced numbers of the nematode significantly.

VI. DISCUSSION AND CONCLUSIONS

A number of fungi are now known to be consistently associated with cysts and eggs of plant-parasitic nematodes, and to successfully exploit this ecological niche as a food source. Bearing in mind the hundreds of fungi known to be ubiquitous in most agricultural soils, those regularly involved as colonizers of nematode cysts and eggs are remarkably restricted in number. Whether or not the consistent occurrence of particular fungi in this microhabitat reflects a high degree of specialization or well developed, but unspecialized, opportunistic competitive capacity has yet to be elucidated. There is good reason to believe that such fungi play a significant role in regulating nematode population dynamics. The types of relationships between members of this opportunistic soil mycoflora and their nematode hosts appear to be complex and variable. Much remains to be learned concerning modes of action and microenvironmental conditions that regulate pathogenic activity. Some of the fungi encountered, such as species of *Verticillium*, clearly have some measure of specialization, enzymatic or otherwise, that renders them capable of colonizing and exploiting what is a relatively abundant nutrient source in many soils. In different circumstances, different fungi predominate and geographical location, environmental conditions, soil type, cropping history, and other parameters are thought to be responsible for this.

Fungal populations fluctuate substantially with amounts of available organic matter in soil and are affected by organic and inorganic fertilization and by introduction of other soil additives, including nematicides. An extant mycoflora might well be inadvertently altered through agricultural practice and this may account for the differences known to sometimes occur in levels of suppressiveness. Opportunistic colonizers of nematode cysts and eggs are, however, well adapted to their environment, being mostly successful competitors that are able to survive host-free periods. As more knowledge is gained concerning the role of these fungi in nematode pathology, a greater awareness of the interrelationship between their activity and nematode population levels will ensue.

REFERENCES

1. **Thorne, G.**, *Principles of Nematology*, McGraw-Hill, New York, 1961, 279.
2. **Kühn, J.**, Vorläufiger Bericht über die bisherigen Ergebnisse der seit dem Jahre 1875 im Auftrage des Vereins für Rubenzucker-Industrie ausgeführten Versuche zur Ermittelung der Ursache der Rübenmüdigkeit des Bodens und zur Erforschung der Natur der Nematoden, *Z. Ver. Rubenzucker-Ind. Dtsch. Reich* (ohne Band), 452, 1877.
3. **Kühn, J.**, Die Ergebnisse der Versuche zur Ermittelung der Ursache der Rübenmüdigkeit und zur Erforschung der Natur der Nematoden, *Ber. Physiol. Lab. Versuchsanst. Landwirthsch. Inst. Univ. Halle*, 3, 1, 1881.
4. **Hollrung, D.**, Die Hallesche Versuchsstation für Nematodenvertilgung. Mittheilung über die Arbeiten im Jahre 1889, *Z. Ver. Rubenzucker-Ind. Dtsch. Reich*, 40, 140, 1890.
5. **Baunacke, W.**, Untersuchungen zur Biologie und Bekämpfung des Rübennematoden *Heterodera schachtii* Schmidt, *Arb. Biol. Reichsanst. Land Fortwirtsch, Berlin*, 11, 185, 1922.
6. **Dollfus, R. P.**, *Parasites des Helminthes*, Lechevalier, Paris, 1946, 482.
7. **Korab, J. J.**, Results of a study of the beet nematode *Heterodera schachtii* at the laboratory of the Belaya Tserkov Research Station, (in Russian with English summary), *Sb. Sort. Semenovod. Uprav.*, 8, 29, 1926.
8. **Tribe, H. T.**, On 'Olpidium nematodeae Skvortzow', *Trans. Br. Mycol. Soc.*, 69, 509, 1977.
9. **Tribe, H. T.**, Pathology of cyst-nematodes, *Biol. Rev.*, 52, 477, 1977.
10. **Goffart, H.**, Untersuchungen am Hafernematoden *Heterodera schactii* Schm. unter besonderer Berücksichtigung der schleswig-holsteinischen Verhältnisse, III, *Arb. Biol. Reichsanst. Land Forstwirtsch. Berlin-Dahlem*, 20, 1, 1932.
11. **Rademacher, B. and Schmidt, O.**, Die bisherigen Erfahrungen in der Bekämpfung des Rübennematoden (*Heterodera schachtii* Schm.) auf dem Wege der Reizbeeinflussung, *Arch. Pflanzenbau Berlin*, 10, 237, 1933.
12. **Rozsypal, J.**, Houby na hád' átku fepném *Heterodera schachtii* Schmidt v moravských půdách, *Vestn. Cesk. Akad. Zemed.*, 10, 413, 1934.
13. **Schol-Schwarz, M. B.**, Revision of the genus *Phialophora* (Moniliales), *Persoonia*, 6, 59, 1970.
14. **Van Beyma, F. H.**, Beschreibung einiger neuer Pilzarten aus dem Centraalbureau voor Schimmelcultures, Baarn (Holland), IV. Mitteilung, *Zentbl. Bakt. Parasitkde (Abt. II)*, 96, 411, 1937.
15. **Van Beyma, F. H.**, Beschreibung der im Centraalbureau voor Schimmelcultures vorhandenen Arten der Gattungen *Phialophora* Thaxter und *Margarinomyces* Laxa, nebst Schussel zur ihrer Bestimmung, *Antoine van Leeuwenhoek*, 9, 51, 1943.
16. **Jones, F. G. W.**, Soil populations of the beet eelworm (*Heterodera schachtii* Schm.) in relation to cropping, *Ann. Appl. Biol.*, 32, 351, 1945.
17. **Triffit, M. J.**, On cyst-like bodies, resembling cysts of *Heterodera schachtii* of common occurrence in British soils, *J. Helminthol.*, 13, 59, 1935.
18. **Gerdemann, J. W. and Nicholson, T. H.**, Spores of mycorrhizal *Endogone* species extracted from soil by wet sieving and decanting, *Trans. Br. Mycol. Soc.*, 46, 235, 1963.
19. **Van der Laan, P. A.**, Een Schimmel asl parasiet van de cyste-inhoud van de aardappelcystenaaltje (*Heterodera rostochiensis* Wollenw.), *Tijdschr. Plantenziekten*, 59, 101, 1953.
20. **Van der Laan, P. A.**, Onderzoekingen over schimmels, die parasiteren op de cyste-inhoud van het aardappelcystennaltje (*Heterodera rostochiensis* Wollenw.), *Tijdschr. Plantenziekten*, 62, 305, 1956.
21. **Willcox, J. and Tribe, H. T.**, Fungal parasitism in cysts of *Heterodera*. I. Preliminary investigations, *Trans. Br. Mycol. Soc.*, 62, 585, 1974.
22. **Bursnall, L. A. and Tribe, H. T.**, Fungal parasitism in cysts of *Heterodera*. II. Egg parasitism of *H. schachtii*, *Trans. Br. Mycol. Soc.*, 62, 595, 1974.
23. **Graham, C. W. and Stone, L. E. W.**, Field experiments on the cereal cyst-nematode (*Heterodera avenae*) in south-east England 1967-72, *Ann. Appl. Biol.*, 80, 61, 1975.
24. **Kerry, B. R.**, Fungi and the decrease of cereal cyst-nematode populations in cereal monoculture, *EPPO Bull.*, 5, 353, 1975.
25. **Kerry, B. R. and Crump, D. H.**, Observations on fungal parasites of females and eggs of the cereal cyst-nematode, *Heterodera avenae*, and other cyst nematodes, *Nematologica*, 23, 193, 1977.
26. **Goswami, B. K. and Rumpenhorst, H. J.**, Association of an unknown fungus with potato cyst nematodes, *Globodera rostochiensis* and *G. pallida*, *Nematologica*, 24, 251, 1978.
27. **Tribe, H. T.**, Extent of disease in populations of *Heterodera*, with especial reference to *H. schachtii*, *Ann. Appl. Biol.*, 92, 61, 1979.
28. **Nigh, E. A., Thomason, I. J., and van Gundy, S. D.**, Identification and distribution of fungal parasites of *Heterodera schachtii* eggs in California, U.S.A., *Phytopathology*, 70, 884, 1980.
29. **Morgan-Jones, G. and Rodríguez-Kábana, R.**, Fungi associated with cysts of *Heterodera glycines* in an Alabama soil, *Nematropica*, 11, 69, 1981.

30. **Morgan-Jones, G., Gintis, B. O., and Rodríguez-Kábana, R.,** Fungal colonization of *Heterodera glycines* cysts in Arkansas, Florida, Mississippi and Missouri soils, *Nematropica,* 11, 155, 1981.

31. **Gintis, B. O., Morgan-Jones, G., and Rodriguez-Kabana, R.,** Mycoflora of young cysts of *Heterodera glycines* in North Carolina soils, *Nematropica,* 12, 295, 1982.

32. **Stirling, G. R. and Kerry, B. R.,** Antagonists of the cereal cyst nematode, *Heterodera avenae* Woll. in Australian soils, *Aust. J. Exp. Agric. Anim. Husb.,* 23, 318, 1983.

33. **Kerry, B. R., Crump, D. H., and Mullen, L. A.,** Natural control of the cereal cyst nematode, *Heterodera avenae* Woll., by soil fungi at three sites, *Crop Prot.,* 1, 99, 1982.

34. **Gintis, B. O., Morgan-Jones, G., and Rodríguez-Kábana, R.,** Fungi associated with several developmental stages of *Heterodera glycines* from an Alabama soybean field soil, *Nematropica,* 13, 181, 1983.

35. **Dunn, M. T., Sayre, R. M., Carrell, A., and Wergin, W. P.,** Colonization of nematode eggs by *Paecilomyces lilacinus* (Thom) Samson as observed with scanning electron microscope, *Scanning Electron Microsc.,* 3, 1351, 1982.

36. **Dunn, M. T.,** *Paecilomyces nostocioides,* a new hyphomycete isolated from cysts of *Heterodera zeae, Mycologia,* 75, 179, 1983.

37. **Godoy, G., Rodríguez-Kábana, R., and Morgan-Jones, G.,** Fungal parasites of *Meloidogyne arenaria* eggs in an Alabama soil. A mycological survey and greenhouse studies, *Nematropica,* 13, 201, 1983.

38. **Morgan-Jones, G., Rodríguez-Kábana, R., and Tovar, J. G.,** Fungi associated with cysts of *Heterodera glycines* in the Cauca Valley, Colombia, *Nematropica,* 14, 173, 1984.

39. **Tovar, J. G. and Medina, C.,** *Heterodera glycines* en soya y frijol en el Valle del Cauca, Colombia, *Nematropica,* 13, 229, 1983.

40. **Dackman, C. and Nordbring-Hertz, B.,** Fungal parasites of the cereal cyst nematode *Heterodera avenae* in southern Sweden, *J. Nematol.,* 17, 50, 1985.

41. **Morgan-Jones, G. and Rodríguez-Kábana, R.,** Fungi associated with cysts of potato cyst nematodes in Peru, *Nematropica,* 16, 21, 1986.

42. **Linford, M. B. and Oliveira, J. M.,** Potential agents of biological control of plant parasitic nematodes, *Phytopathology,* 28, 14, 1938.

43. **Stirling, G. R. and Mankau, R.,** *Dactylella oviparasitica,* a new fungal parasite of *Meloidogyne* eggs, *Mycologia,* 70, 774, 1978.

44. **Stirling, G. R. and Mankau, R.,** Mode of parasitism of *Meloidogyne* and other nematode eggs by *Dactylella oviparasitica, J. Nematol.,* 11, 282, 1979.

45. **Jatala, P., Kalchenbach, R., and Bocangel, M.,** Biological control of *Meloidogyne incognita acrita* and *Globodera pallida* on potatoes, *J. Nematol.,* 11, 303, 1979.

46. **Morgan-Jones, G., White, J. F., and Rodríguez-Kábana, R.,** Fungal parasites of *Meloidogyne incognita* in an Alabama soybean field soil, *Nematropica,* 14, 93, 1984.

47. **Lysek, H.,** Effect of certain soil organisms on the eggs of parasitic roundworms, *Nature (London),* 199, 925, 1963.

48. **Lysek, H.,** Study of biology of geohelminths. II. The importance of some soil microorganisms for the viability of geohelminth eggs in soil, *Acta Univ. Palacki. Olomu.,* 40, 83, 1966.

49. **Lysek, H.,** A scanning electron microscope study of the offset of an ovicidal fungus on the eggs of *Ascaris lumbricoides, Parasitology,* 77, 139, 1978.

50. **Barron, G. L.,** *The Nematode-Destroying Fungi,* Canadian Biological Publications, Guelph, 1977, 84.

51. **Tribe, H. T.,** Prospects for the biological control of plant-parasitic nematodes, *Parasitology,* 81, 619, 1980.

52. **Godoy, G., Rodríguez-Kábana, R., and Morgan-Jones, G.,** Parasitism of eggs of *Heterodera glycines* and *Meloidogyne arenaria* by fungi isolated from cysts of *H. glycines, Nematropica,* 12, 111, 1982.

53. **Rodríguez-Kábana, R., Morgan-Jones, G., Dodoy, G., and Gintis, B. O.,** Effectiveness of species of *Gliocladium, Paecilomyces,* and *Verticillium* for control of *Meloidogyne arenaria* in field soil, *Nematropica,* 14, 155, 1984.

54. **Morgan-Jones, G., White, J. F., and Rodríguez-Kábana, R.,** Phytonematode pathology: ultrastructural studies. I. Parasitism of *Meloidogyne arenaria* eggs by *Verticillium chlamydosporium, Nematropica,* 13, 245, 1983.

55. **Morgan-Jones, G., White, J. F., and Rodríguez-Kábana, R.,** Phytonematode pathology: ultrastructural studies. II. Parasitism of *Meloidogyne arenaria* eggs and larvae by *Paecilomyces lilacinus, Nematropica,* 14, 57, 1984.

56. **Morgan-Jones, G. and Rodríguez-Kábana, R.,** Phytonematode pathology: fungal modes of action. A perspective, *Nematropica,* 15, 107, 1985.

57. **Bird, A. F. and McClure, M. A.,** The tylenchid (Nematoda) egg shell: structure, composition and permeability *Parasitology,* 72, 19, 1976.

58. **Clarke, A. J. and Shepherd, A. M.,** Hatching factors for the potato cyst nematode, *Heterodera rostochiensis* Woll., *Ann. Appl. Biol.,* 61, 139, 1968.

59. **Bird, A. F.,** *The Structure of Nematodes*, Academic Press, New York, 1971, 295.
60. **Booth, C.,** The genus *Cylindrocarpon, Mycol. Pap.,* 104, 1, 1966.
61. **Hams, A. F. and Wilkin, G. D.,** Observations on the use of predacious fungi for control of *Heterodera* spp., *Ann. Appl. Biol.,* 49, 515, 1961.
62. **Taylor, G. S. and Parkinson, D.,** Studies on fungi in the root region. IV. Fungi associated with the roots of *Phaseolus vulgaris* L., *Plant Soil,* 22, 1, 1965.
63. **Fijan, N.,** Systemic mycosis in channel catfish, *Bull. Wildlife Dis. Assoc.,* 5, 109, 1969.
64. **De Hoog, G. S. and Hermanides-Nijhof, E. J.,** The black yeasts and allied hyphomycetes, *Stud. Mycol.,* 15, 1, 1977.
65. **Domsch, K. H. and Gams, W.,** *Fungi in Agricultural Soils*, Longman, London, 1972.
66. **Vesonder, R. F. and Hesseltine, C. W.,** Metabolites of *Fusarium,* in *Fusarium: Diseases, Biology, and Taxonomy,* Nelson, P. E., Toussoun, T. A., and Cook, R. K., Eds., Pennsylvania State University Press, University Park, Pa., chap. 32.
67. **Domsch, K. H., Gams, W., and Anderson, T.-H.,** *Compendium of Soil Fungi*, Vol. 1, Academic Press, London, 1980, 530.
68. **Samson, R. A.,** *Paecilomyces* and some allied hyphomycetes, *Stud. Mycol.,* 6, 1, 1974.
69. **Murao, S., Yamamoto, R., and Arai, M.,** Isolation and identification of red yeast cell wall lytic enzyme producing microorganism, *Agric. Biol. Chem.,* 40, 23, 1976.
70. **Oshima, S. and Akazawa, Y.,** Pathogenicity of *Paecilomyces lilacinus* (Thom) Samson to potato tuber moth and its applications to control of the insect, *Bull. Okayama Tob. Exp. Stn.,* 41, 73, 1980.
71. **Agrawal, P. K., Lal, B., Wahab, S., Srivastava, O. P., and Misra, S. C.,** Orbital paecilomycosis due to *Paecilomyces lilacinus* (Thom) Samson, *Saboraudia,* 17, 363, 1979.
72. **Takayasu, S., Akagi, M., and Shimizu, Y.,** Cutaneous mycosis caused by *Paecilomyces lilacinus, Arch. Dermatol.,* 113, 1687, 1977.
73. **Jatala, P., Kaltenbach, R., Bocangel, M., Devaux, A. J., and Campos, R.,** Field application of *Paecilomyces lilacinus* for controlling *Meloidogyne incognita* on potatoes, *J. Nematol.,* 12, 226, 1980.
74. **Jatala, P., Salas, R., Kaltenbach, R., and Bocangel, M.,** Multiple application and long-term effect of *Paecilomyces lilacinus* in controlling *Meloidogyne* under field conditions, *J. Nematol.,* 13, 445, 1981.
75. **Dickson, D. W. and Mitchell, D. J.,** Evaluation of *Paecilomyces lilacinus* as a biocontrol agent of *Meloidogyne javanica* on tobacco, *J. Nematol.,* 17, 519, 1985.
76. **Culbreath, A. K., Rodríguez-Kábana, R., and Morgan-Jones, G.,** Chitin and *Paecilomyces lilacinus* for control of *Meloidogyne arenaria, Nematropica,* 16, 153, 1986.
77. **Kerry, B. R., Simon, A., and Rovira, A. D.,** Observations on the introduction of *Verticillium chlamydosporium* and other parasitic fungi into soil for control of the cereal cyst-nematode *Heterodera avenae, Ann. Appl. Biol.,* 105, 509, 1984.
78. **Morgan-Jones, G. and Rodríguez-Kábana, R.,** unpublished data.
79. **Morgan-Jones, G. and Rodríguez-Kábana, R.,** unpublished data.

Chapter 3

INFECTION EVENTS IN THE FUNGUS-NEMATODE SYSTEM

Hans-Börje Jansson and Birgit Nordbring-Hertz

TABLE OF CONTENTS

I. INTRODUCTION

Infection of vermiform nematodes by nematophagous fungi generally follows the same route whether the nematode is captured in adhesive traps of predatory fungi or mediated by conidia of endoparasites. A successful infection always results in complete digestion of the nematode corpus. The total time for the infection process varies between fungal species and also between species of nematodes being parasitized. The zoosporic phycomycete *Catenaria anguillulae* usually completes its life cycle in a rhabditoid nematode in 24 hr, whereas the hyphomycetous endoparasite *Meria (Drechmeria) coniospora* takes up to 72 hr for the same process. The infection of nematodes by nematophagous fungi usually follows a general sequence of events starting with recognition between the two types of organisms. We will use the term recognition in a broad sense including attraction phenomena as well as inter-actions on a molecular level involved in adhesion.[1,2] The recognition step is followed by penetration of the nematode cuticle, possibly a toxification step, and finally the animal is digested by trophic hyphae of the fungus. The different infection events, some of which may not necessarily be present in all fungal-nematode interactions, will be discussed in this chapter.

What is striking in their interactions with nematophagous fungi is the behavioral pattern of the nematodes which contribute directly to their own destruction. Firstly, the nematodes themselves induce the formation of fungal trapping organs in which they are later captured.[3] Secondly, the nematodes are generally attracted to their fungal enemies (see below); and thirdly, nematophagous fungi apparently capture and penetrate the cuticle only of live ne-matodes, with dead animals being invaded via the natural orifices (see below). This strange behavior of the nematodes may lead us to speculate that the nematophagous fungi in many cases mimic a food source or a host of a parasitic nematode.

II. ATTRACTION

An important way for organisms to facilitate contact is by moving towards each other. This attraction can be mediated by chemical substances or have other causes. In a host-parasite relationship the parasite is usually attracted to its host, but the opposite behavior is also possible. Attraction phenomena involved in the interactions between nematodes and nematophagous fungi can be divided into two categories: active movement (chemotaxis and chemotropism) of fungi to the nematodes, and attraction (chemotaxis) of nematodes to their fungal parasites or predators.

From the first category very little experimental evidence exists. Zoospores of fungi such as *Catenaria anguillulae* often accumulate at the body orifices of nematodes and it was therefore suggested that the zoospores were attracted to leaking substances from the mouth, anus, etc.[4] Whether zoospores of fungi attacking females of cyst nematodes, e.g., *Nema-tophthora gynophila*, are chemically attracted to their hosts is so far unknown. A chemotactic behavior of the zoosporic nematophagous fungi may be important in the infection process, as has been generally recognized for plant parasitic zoosporic fungi, e.g., the *Phytophthora* spp. and the *Pythium* spp.[5]

Chemotropic behavior, i.e., directed growth of hyphae, has been observed in the nematode egg-parasitizing fungus *Rhopalomyces elegans*, which apparently grows towards exudates of nematode eggs.[4] Chemotropism toward dead nematodes was also observed for hyphae growing out from detached constricting ring traps of *Dactylella doedycoides*.[6] Generally, very little is known about fungal chemotaxis and chemotropism in relation to nematodes and more attention should be paid to this interesting question.

More work has been done on attraction of nematodes to nematophagous fungi. It appears that nematophagous fungi generally attract their prey.[7-11] In only one instance has a nematode

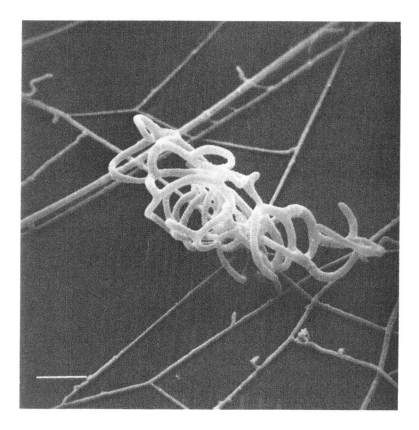

FIGURE 1.   SEM micrograph of adhesive network trap of *Arthrobotrys oligospora*. Bar: 20 μm. (From Nordbring-Hertz, B., *Appl. Environ. Microbiol.*, 45, 290, 1983. With permission.)

species, *Panagrellus redivivus*, been reported as being repelled by a species, *Arthrobotrys arthrobotryoides*, and in some cases the response was neutral.[7] As an overall measure, about 75% of the nematophagous fungi examined attracted several different nematode species.[8]

Since the group of nematophagous fungi covers species from facultative saprophytes to obligate parasites with varying food preferences, the ability to attract nematodes may be important to their survival in soil. In laboratory studies we found that endoparasitic fungi, e.g., *Harposporium anguillulae*, showed a higher ability to attract nematodes than the more saprophytic species, e.g., *Arthrobotrys oligospora*.[7] This effect was strongly correlated with the ability of the fungi to destroy nematodes (predacity) when measured in soil microcosm experiments.[12]

Many of the predatory fungi can grow in a saprophytic phase without traps, or in a predacious phase with trapping organs. A typical adhesive network trap of *A. oligospora*[13] is shown in Figure 1. The presence of traps has a marked influence on the ability of the fungus to attract nematodes, the attraction increasing by a factor of two after induction of trap formation[12] (Figure 2). The endoparasitic fungi use their spores to attack nematodes, and also, in this case, the infecting structures are important for attraction of nematodes, as shown for adhesive conidia of *M. coniospora* (Figure 3) and other fungi.[14]

Since nematodes appear to be attracted to many different substances[15] it is possible that part of the attraction to nematophagous fungi is due to CO_2 or other common metabolic products, and partly due to more species-specific attractants. The attractive compounds have not been identified, but seem to be of both volatile and nonvolatile nature.[7]

In one species among the endoparasitic fungi, *M. coniospora*, the infective conidia attach

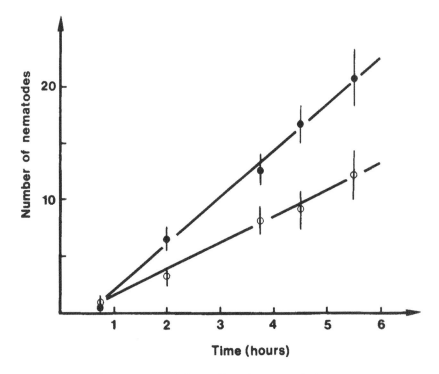

FIGURE 2.   Attraction of the nematode *Panagrellus redivivus* to *Arthrobotrys oligospora* mycelium with traps (●), and mycelium without traps (○). The steeper the slope of the line, the higher is the ability of the fungus to attract nematodes. (From Jansson, H-B., *Microb. Ecol.*, 8, 233, 1982. With permission.)

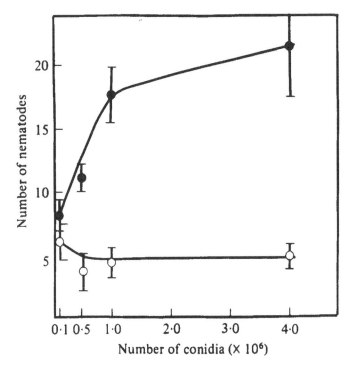

FIGURE 3.   Attraction of *Panagrellus redivivus* to increasing numbers of conidia of *Meria coniospora* (●), and controls without conidia (○). (From Jansson, H-B., *Trans. Br. Mycol. Soc.*, 79, 25, 1982. With permission.)

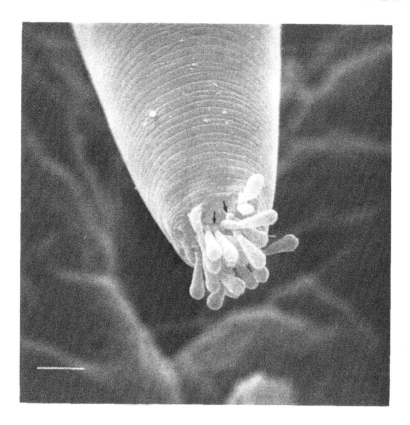

FIGURE 4.   SEM micrograph of conidia of *Meria coniospora* adhering to sensory structures in cephalic region of *Panagrellus redivivus*. Note adhesive buds of the conidia (arrows). Bar: 5 μm. (From Jansson, H-B., *Trans. Br. Mycol. Soc.*, 79, 25, 1982. With permission.)

to the head and tail region of most nematodes tested.[16,17] The specific sites of attachment were the chemosensory structures of the nematodes (Figure 4). After adhesion of the spores the nematodes lost their ability to respond chemotactically to several sources of attractants.[17] The bacterial-feeding nematodes, *Caenorhabditis elegans* and *P. redivivus*, apparently have chemoreceptors consisting of glycoproteins, where the terminal carbohydrates, sialic acid and mannose, appear to be vital for chemotaxis of these nematodes to bacterial filtrates and to nematophagous fungi.[18-20] Furthermore, sialic acid seems to be important also in the adhesion of conidia (see next section). Thus, it appears that in the interactions between *M. coniospora* and nematodes there is a connection between attraction and adhesion.

III. ADHESION

The connection between attraction and adhesion in recognition of prey by the predatory fungi is not entirely obvious. However, as mentioned above, the attraction of nematodes to mycelia with adhesive traps was significantly higher than to hyphae without traps. Because of this, and the observation of the general behavior of nematodes in the vicinity of the fungi, chemoattraction is considered as a first step in the recognition process followed by the firm adhesion of the nematodes to adhesive traps. The adhesive material of a few fungi, e.g., the *Nematoctonus* spp. and the *Stylopage* spp., can easily be observed directly under the light microscope.[4] However, in most nematophagous fungi the adhesive can only be observed with the electron microscope.[21-26] In *A. oligospora* the adhesion process has been studied in detail and the work on this fungus will be reported below.

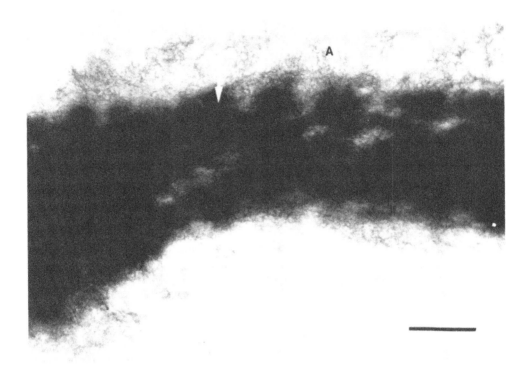

FIGURE 5.   TEM micrograph showing adhesive (A) of trap cell of *Arthrobotrys oligospora*. Trap cell is recognized by the presence of dense bodies (arrow). Bar: 1 μm. (From Veenhuis, M., Nordbring-Hertz, B., and Harder, W., *Antonie van Leeuwenhoek*, 51, 385, 1985. With permission.)

Adhesive material is present on the surface of the traps of *A. oligospora* even before the interaction between fungus and nematode occurs[21,24,27] (Figure 5). At the contact between trap and nematode an increased secretion of adhesive takes place and probably concomitantly, a change in adhesive properties occurs.[21,24] The fibrils of the adhesive become directed perpendicularly to the host surface, probably to facilitate the anchoring of the nematode to the trap and the further fungal invasion of the nematode[24] (Figure 6). A different type of adhesive was observed in *M. coniospora* (Figure 7), and this seems to be composed of radiating fibrils irrespective of contact with a nematode or not. Saikawa[25] suggested that this adhesive was glycocalyx-like and contained polysaccharides.

In *A. oligospora* and some other fungi there is now evidence for a molecular interaction between a carbohydrate-binding protein located on the trapping structure binding to a carbohydrate on the surface of the nematode[17,28-33] (Table 1).

This lectin-carbohydrate interaction was first studied in inhibition experiments using intact organisms (nematodes and fungi). In such hapten inhibition experiments the lectin-bearing structure was subjected to the carbohydrate involved. It became clear that the adhesion of nematodes to the traps of *A. oligospora* was inhibited by N-acetylgalactosamine (GalNAc).[28,31] Capture of nematodes by other nematode-trapping species were inhibited by other sugars,[30] whereas in the only endoparasite investigated, *M. coniospora*, inhibition occurred only by sialic acid.[17] Thus, the specificities of the proposed lectins of nematophagous fungi vary considerably, indicating the presence of several sugar residues on the nematode surfaces investigated and/or different carbohydrate-binding proteins on the fungal structures. Furthermore, in all cases studied, treatment of the lectin-bearing structure with trypsin or glutaraldehyde also inhibited capture, indicating the protein nature of the binding molecule.[18,30,33]

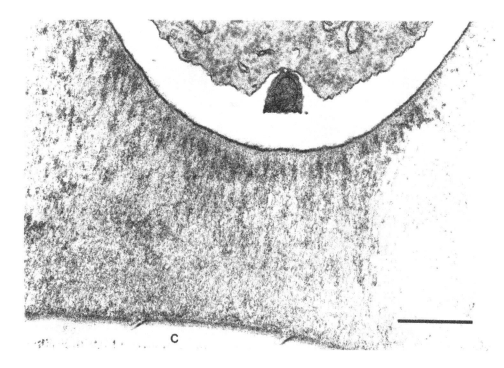

FIGURE 6. TEM of adhesive of *Arthrobotrys oligospora*. After capture of a nematode the fibrils of the adhesive are oriented in one direction at the site of capture. (C) nematode cuticle. Bar: 0.5 μm. (From Veenhuis, M., Nordbring-Hertz, B., and Harder, W., *Antonie van Leeuwenhoek*, 51, 385, 1985. With permission.)

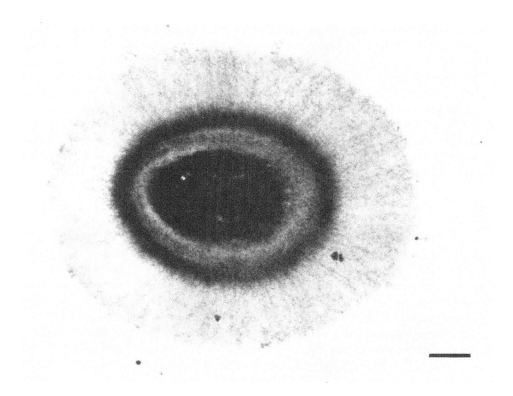

FIGURE 7. TEM of adhesive bud of *Meria coniospora* conidium showing fibrillar adhesive coating. Bar: 0.1 μm. Courtesy of Dr. M. Veenhuis.

Table 1
**NEMATOPHAGOUS FUNGI SUGGESTED TO
PRODUCE LECTINS ON THEIR INFECTIVE
STRUCTURES AND CORRESPONDING LECTIN-
BINDING CARBOHYDRATE ON THE NEMATODE
CUTICLE**

Fungus	Lectin-binding carbohydrate	Ref.
Arthrobotrys oligospora	N-acetyl-D-galactosamine	28
A. conoides	D-glucose/D-mannose	30
Dactylaria candida	2-Deoxy-D-glucose	33
Meria coniospora	Sialic acid	17
Monacrosporium eudermatum	L-fucose	30
M. rutgeriensis	2-Deoxy-D-glucose	30

It is important to stress that in the nematophagous fungi studied only a certain developmental phase of the fungus (the trap or the spore) is responsible for the capture of nematodes. In many cases this phase is adhesive. The question now is whether this adhesive is to be considered equivalent to the lectin. To solve this problem a series of investigations aiming at localizing the proposed lectin have been performed during the last few years. Traps of *A. oligospora* were produced in comparatively large amounts in a liquid culture procedure.[34] The trap-containing mycelium was surface labelled with (^{125}I)-iodosulfanilic acid and then homogenized. In two affinity chromatography steps using GalNAc-Sepharose and a metal chelate affinity column, a GalNAc-specific, Ca^{2+}-binding protein was isolated.[29] This protein had a molecular weight of about 20,000 daltons, and it was not present on mycelia without traps.[35]

To investigate whether this protein was located in the adhesive, on the surface of the trap or within the cytoplasm of the trap, antibodies to the protein were raised in a rabbit.[36] Using an immunocytochemical method with the protein A/gold technique sections of traps, hyphae, and traps invading nematodes were investigated in the electron microscope. Gold particles were found almost exclusively in the trap cell wall, very little in hyphal cell walls, and not at all in the adhesive.[37] Thus the lectin seems indeed to be present on the structure responsible for capture. Its function in starting the infection process has still to be elucidated. However, the hypothesis that the binding of a carbohydrate on the nematode surface to a trap lectin initiates the further interaction between the organisms — cuticle penetration and digestion — is still valid.[28,38]

IV. CUTICLE PENETRATION

Over the years many light and electron microscopic studies illustrating the penetration of nematode cuticles by different nematophagous fungi have been published.[21-26,39-45] It seems clear that cuticles of living nematodes can only be penetrated after the firm anchoring of the infection structures to the nematode surface either by an adhesive[21-24,46] or mechanically as in the constricting ring.[39,42] The mechanism involved in the penetration of the cuticle has been connected with certain electron dense organelles present in both adhesive and mechanical trapping structures (Figure 8). These organelles, not present in ordinary hyphae, have been suggested to be responsible either for the release of adhesive or enzymes involved in penetration, or the hydrolytic enzymes functioning in digestion of the nematode.[21-23,41,42,45] One reason for suggesting that the electron dense bodies were involved in either trapping or penetration was the observation that the organelles disappeared on invasion of the nematodes.[21,41,44] However, in recent studies dense bodies have been detected in the infection

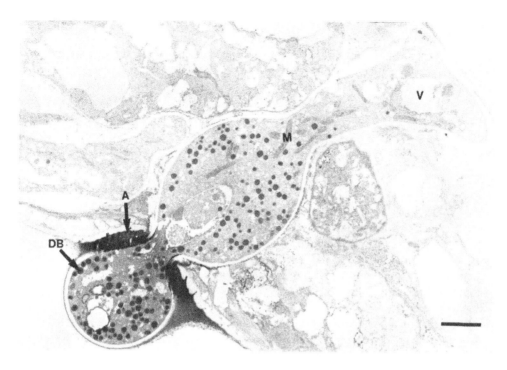

FIGURE 8. TEM of penetration of nematode cuticle by *Arthrobotrys oligospora*. Note electron dense bodies (DB), adhesive coating (A), mitochondria (M), and vacuoles (V). The adhesive coating, trap cell wall and cell membrane show staining for acid phosphatase. Bar: 1 μm. (From Veenhuis, M., Nordbring-Hertz, B., and Harder, W., *Antonie van Leeuwenhoek*, 51, 385, 1985. With permission.)

bulb shortly after penetration of the nematode[23,24,43] (Figure 8). The differences in these results are probably due to the fact that time-related observations have been reported only in a few cases.[21,23,24,47]

Using a combination of different light and electron microscopic techniques the time sequence of the capture, penetration, and digestion of nematodes by *A. oligospora* was studied in a film.[47] Here it was confirmed that the penetration of *P. redivivus* took place within an hour[21] and often after only 20 to 30 min. Dense bodies were transported from the trap and appeared in the infection bulb[24,43] during the first hour of penetration and then disappeared.[47] These results suggest that the dense bodies do not participate either in secretion of adhesive or in lysis of the cuticle.

Cytochemical studies of the interaction between *A. oligospora* and *P. redivivus* have shown clearly that the dense bodies contain catalase and D-amino acid oxidase[43] and thus are peroxisomal in nature. They do not carry hydrolytic enzymes.[24,46] Acid phosphatase activity was only detected in the adhesive layer invariably present between trap and nematode, and later in infection bulbs and trophic hyphae within the invaded nematode, suggesting that the fungus actively degrades its prey.

Whether penetration of the cuticle is enzymatic or mechanical or both is not yet clear. It is interesting to note that normal hyphae penetrate the nematode only through its body orifices,[21] whereas hyphae originating from traps which have a high metabolic activity in comparison to normal hyphae[3,23] are able to penetrate the cuticle. Veenhuis et al.[24] suggested that a prerequisite for the fungus to penetrate the nematode against the osmotic pressure of the intact animal is the firm anchoring by the adhesive. Observations of adhesive properties and ultrastructural-cytochemical studies of the interaction events[24,46,47] made these authors suggest that the penetration occurs, at least largely, by mechanical force. Indentations in and stretching of the cuticle before the penetration, without signs of hydrolysis of the cuticle, strengthen this point of view.

It is a striking feature that nematophagous fungi penetrate the cuticle of living nematodes and only some time after penetration the nematodes become moribund. The length of this time depends on the fungal species and the type of nematode. In *A. oligospora — P. redivivus* the nematode dies soon after the formation of the first trophic hyphae;[47] whereas in third stage larvae of some animal parasitic nematodes, attacked by *M. coniospora*, the cuticles were not penetrated at all until the protective sheath, consisting of the cast cuticle from the second molt, was removed and colonization could take place.[16]

V. TOXIFICATION

When observing nematodes and nematophagous fungi under the microscope on agar plates one is often amazed by the rapidity with which nematodes are immobilized after capture by some nematophagous fungi, e.g., *A. oligospora*. The movement of the nematodes is often arrested within only 10 to 15 min. The nematode is apparently not dead since addition of a drop of water usually revitilizes the captured nematode. With other fungi, e.g., the endoparasite *M. coniospora*, the nematode remains active and motile for more than 24 hr with the fungus growing inside the nematode.[26] The possible production of nematicidal toxins has intrigued many scientists, but very little information has been published.

Olthof and Estey[48] were using *A. oligospora* to test toxicity to a *Rhabditis* sp. They grew the fungus in liquid culture and treated the nematodes in three different ways: with sterile extracts of the fungus alone, with filtrate from crushed nematodes, and with filtrate from crushed nematodes which had been parasitized by the fungus. Only the nematodes treated with the filtrate from parasitized worms showed signs of toxicity and the authors drew the conclusion that a fungal toxin was produced following capture of the nematodes.

Balan and Gerber[10] showed that *A. dactyloides* grown in liquid culture produced a toxin in the culture filtrate which killed *P. redivivus*, and that maximum nematicidal activity coincided with production of ammonia in the broth. The authors suggested that ammonia may be the toxin produced by the fungus. Later, culture filtrates from three other predatory fungi were shown to produce nematicidal activity irrespective of addition of nematodes to the fungal culture.[49] Since nematicidal activity has also been shown from cultures of saprophytic as well as entomophagous fungi[50,51] it is doubtful whether the above mentioned immobilization of nematodes is due to substances released into liquid medium by the predatory fungi. Probably a toxic agent has to be "injected" into the captured nematode.

Endoparasitic species within the genus *Nematoctonus* appear to produce toxins which kill nematodes after adhesion of the fungus but before visible signs of penetration.[52] Both crushed mycelial extracts and culture filtrates of *N. haptocladus*, *N. concurrens*, and *N. robustus* showed nematotoxic effects, whereas such material from other fungi did not.[52,53] Giuma et al.[54] suggested that the toxin was a thermostable, high molecular weight carbohydrate-containing substance produced by the fungi and released into the medium. They further suggested that the toxin was acting unselectively regarding nematode species, since a mixture of nematodes recovered directly from soil were killed by the fungal extracts in the same way as their test nematode, *Aphelenchus avenae*.

Table 2 shows a summary of results obtained from tests for production of nematicidal compounds produced by nematophagous fungi. As shown, results vary even for the same fungal species; for instance, culture filtrates of *A. oligospora* had nematotoxic effects in one experiment[49] but not in two other investigations.[48,53] Similarly confusing results were obtained with culture filtrates from *N. haptocladus*. It seems that the toxic substances obtained in culture filtrates may be dependent on growth media and the culture condition of the fungi, and may not be the same compound(s) as the one(s) involved in immobilization of nematodes

Table 2
PRODUCTION OF NEMATICIDAL COMPOUNDS BY NEMATOPHAGOUS FUNGI

Fungal species	Tested as	Toxin	Ref.
Arthrobotrys oligospora	Crushed infected nematodes	Yes	48
	Culture filtrate	No	48, 53
	Culture filtrate	Yes	49
A. conoides	Culture filtrate	Yes	49
A. dactyloides	Culture filtrate	Yes	10
A. arthrobotryoides	Culture filtrate	No	53
Dactylella lysipaga	Culture filtrate	No	53
D. pyriformis	Culture filtrate	Yes	55
D. thaumasia	Culture filtrate	Yes	55
Monacrosporium rutgeriensis	Culture filtrate	Yes	49
M. doedycoides	Culture filtrate	No	53
M. ellipsosporum	Culture filtrate	No	53
Harposporium anguillulae	Culture filtrate	No	53, 54
Nematoctonus haptocladus	Mycelial extract, culture filtrate	Yes	52
	Culture filtrate	No	53
N. concurrens	Mycelial extract, culture filtrate	Yes	52
N. tripolitanus	Culture filtrate	Yes	54
N. robustus	Culture filtrate	No	53

following capture by the fungi in vivo. There is still much more research to be done before the importance of a possible toxin can be evaluated.

VI. DIGESTION

Irrespective of the method of attack by the fungi the nematode corpus is invaded and consumed. In most of the endoparasites the nematode body is in fact the only site where vegetative mycelium can develop. After consumption of the nematode, conidiophores break through the nematode cuticle and conidia are formed in large amounts[26] (Figure 9), and the infection cycle can start again.

The digestion of nematodes by the predatory fungi is often completed within a 48 to 72 hr period. After penetration of the cuticle and formation of an infection bulb, trophic hyphae develop throughout the nematode body. After consumption new loops are formed on the traps and, at the site of the consumed nematode, large numbers of conidiophores and conidia later develop.

How the digestion of the nematode is brought about is not yet fully clear. In *A. oligospora*, cytochemical staining for acid phosphatase at different stages of penetration showed presence of this enzyme not only in the adhesive at the penetration spot but also within the nematode after invasion. Membranous structures of trophic hyphae were heavily stained for acid phosphatase 6 hr after capture. At this stage the nematode cytoplasm also showed heavy staining.[24] The hydrolyzed products apparently are later converted to lipids, since lipid droplets appear in the infection bulb and trophic hyphae a few hours after capture, and predominate there after another 24 hr.[47] Since the dense bodies do not carry hydrolytic enzymes they are not likely to have a function in the digestion of the nematode.[46] The presence of many dense bodies, however, might indicate that β-oxidation of lipids occurs later on during development of new vegetative mycelium.[56]

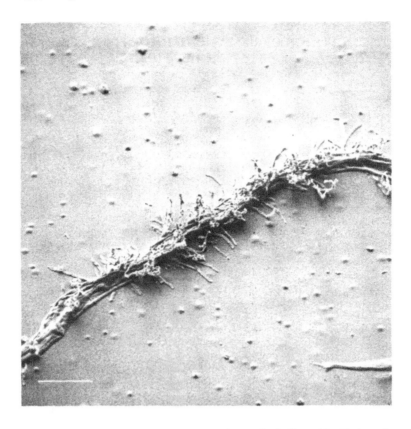

FIGURE 9.   SEM micrograph showing nematode *completely* digested by *Meria conios-pora*. Only protruding conidiophores can be seen outside the nematode corpus. Bar: 10 μm. (From Jansson, H-B., von Hofsten, A., and von Mecklenburg, M., *Antonie van Leeuwenhoek*, 51, 385, 1985. With permission.)

VII. CONCLUSIONS

Why do nematophagous fungi capture, kill, and digest living nematodes? It must obviously be an ecological advantage to the fungi in their competition with other soil microorganisms. Thorn and Barron[57] reported that several wood-decaying fungi, e.g., the oyster mushroom (*Pleurotus ostreatus*), have the ability to capture and consume nematodes. They suggested that the nematodes supplemented the nitrogen diet of the fungi. Some nematode-trapping fungi, mainly the *Arthrobotrys* spp., have the capacity to attack other fungi, i.e., act as mycoparasites.[58,59] The physiological and ecological advantage of this capacity is not yet fully clear, but indicates the diversity of ways in which the nematophagous fungi can cope with varying environmental conditions.

We believe that increased knowledge of the general biology of the nematophagous fungi, together with further studies in soil ecology, may help us to understand how these fascinating fungi behave, and what their functions are in the natural environment.

REFERENCES

1. **Jansson, H-B.**, Receptors and recognition in nematodes, in *Vistas on Nematology*, Veech, J. and Dickson, D., Eds., Society of Nematologists, Tifton, Ga., 1987, 153.
2. **Nordbring-Hertz, B.**, Ecology and recognition in the nematode-nematophagous fungus system, in *Advances in Microbial Ecology*, Vol. 10, Marshall, K. C., Ed., Plenum Press, New York, 1987, 81.
3. **Nordbring-Hertz, B.**, Nematode-induced morphogenesis in the predacious fungus *Arthrobotrys oligospora*, *Nematologica*, 23, 443, 1977.
4. **Barron, G. L.**, *The Nematode-Destroying Fungi, Topics in Mycobiology No. 1*, Canadian Biological Publications, Guelph, Ontario, 1977.
5. **Chet, I. and Mitchell, R.**, Ecological aspects of microbial chemotactic behavior, *Annu. Rev. Microbiol.*, 30, 221, 1976.
6. **Zachariah, K.**, Chemotropism by isolated ring traps of *Dactylella doedycoides*, *Protoplasma*, 106, 173, 1981.
7. **Jansson, H-B. and Nordbring-Hertz, B.**, Attraction of nematodes to living mycelium of nematophagous fungi, *J. Gen. Microbiol.*, 112, 89, 1979.
8. **Jansson, H-B. and Nordbring-Hertz, B.**, Interactions between nematophagous fungi and plant-parasitic nematodes: attraction, induction of trap formation and capture, *Nematologica*, 26, 383, 1980.
9. **Field, J. I. and Webster, J.**, Traps of predacious fungi attract nematodes, *Trans. Br. Mycol. Soc.*, 68, 467, 1977.
10. **Balan, J. and Gerber, N. N.**, Attraction and killing of the nematode *Panagrellus redivivus* by the predacious fungus *Arthrobotrys dactyloides*, *Nematologica*, 18, 163, 1972.
11. **Monoson, H. L., Galsky, A. G., Griffin, J. A., and McGrath, E. J.**, Evidence for and partial characterization of a nematode attraction substance, *Mycologia*, 65, 78, 1973.
12. **Jansson, H-B.**, Predacity by nematophagous fungi and its relation to the attraction of nematodes, *Microb. Ecol.*, 8, 233, 1982.
13. **Nordbring-Hertz, B.**, Dialysis membrane technique for studying microbial interactions, *Appl. Environ. Microbiol.*, 45, 290, 1983.
14. **Jansson, H-B.**, Attraction of nematodes to endoparasitic nematophagous fungi, *Trans. Br. Mycol. Soc.*, 79, 25, 1982.
15. **Zuckerman, B. M. and Jansson, H-B.**, Nematode chemotaxis and possible mechanisms of host/prey recognition, *Annu. Rev. Phytopathol.*, 22, 95, 1984.
16. **Jansson, H-B., Jeyaprakash, A., and Zuckerman, B. M.**, Differential adhesion and infection of nematodes by the endoparasitic fungus *Meria coniospora* (Deuteromycetes), *Appl. Environ. Microbiol.*, 49, 552, 1985.
17. **Jansson, H-B. and Nordbring-Hertz, B.**, The endoparasitic fungus *Meria coniospora* infects nematodes specifically at the chemosensory organs, *J. Gen. Microbiol.*, 129, 1121, 1983.
18. **Jansson, H-B. and Nordbring-Hertz, B.**, Involvement of sialic acid in nematode chemotaxis and infection by an endoparasitic nematophagous fungus, *J. Gen. Microbiol.*, 130, 39, 1984.
19. **Jansson, H-B., Jeyaprakash, A., Damon, R. A., Jr., and Zuckerman, B. M.**, *Caenorhabditis elegans* and *Panagrellus redivivus*: enzyme-mediated modification of chemotaxis, *Exp. Parasitol.*, 58, 270, 1984.
20. **Jeyaprakash, A., Jansson, H-B., Marban-Mendoza, N., and Zuckerman, B. M.**, *Caenorhabditis elegans*: lectin-mediated modification of chemotaxis, *Exp. Parasitol.*, 59, 90, 1985.
21. **Nordbring-Hertz, B. and Stålhammar-Carlemalm, M.**, Capture of nematodes by *Arthrobotrys oligospora*, an electron microscope study, *Can. J. Bot.*, 56, 1297, 1978.
22. **Dowsett, J. A., Reid, J., and Hopkin, A. A.**, Microscopic observations on the trapping of nematodes by the predaceous fungus *Dactylaria cionopaga*, *Can. J. Bot.*, 62, 674, 1984.
23. **Wimble, D. B. and Young, T. W. K.**, Ultrastructure of the infection of nematodes by *Dactylella lysipaga*, *Nova Hedwigia*, 40, 9, 1984.
24. **Veenhuis, M., Nordbring-Hertz, B., and Harder, W.**, An electron microscopical analysis of capture and initial stages of penetration of nematodes by *Arthrobotrys oligospora*, *Antonie van Leeuwenhoek*, 51, 385, 1985.
25. **Saikawa, M.**, An electron microscope study of *Meria coniospora*, an endozoic nematophagous fungus, *Can. J. Bot.*, 60, 2019, 1982.
26. **Jansson, H-B., von Hofsten, A., and von Mecklenburg, M.**, Life cycle of the endoparasitic nematophagous fungus *Meria coniospora*: a light and electron microscopic study, *Antonie van Leeuwenhoek*, 50, 321, 1984.
27. **Nordbring-Hertz, B.**, Scanning electron microscopy of the nematode-trapping organs in *Arthrobotrys oligospora*, *Physiol. Plant.*, 26, 279, 1972.
28. **Nordbring-Hertz, B. and Mattiasson, B.**, Action of a nematode-trapping fungus shows lectin-mediated host-microorganism interaction, *Nature (London)*, 281, 477, 1979.

29. **Borrebaeck, C. A. K., Mattiasson, B., and Nordbring-Hertz, B.,** Isolation and partial characterization of a carbohydrate-binding protein from a nematode-trapping fungus, *J. Bacteriol.,* 159, 53, 1984.

30. **Rosenzweig, W. D. and Ackroyd, D.,** Binding characteristics of lectins involved in the trapping of nematodes by fungi, *Appl. Environ. Microbiol.,* 46, 1093, 1983.

31. **Premachandran, D. and Pramer, D.,** Role of N-acetylgalactosamine-specific protein in trapping of nematodes by *Arthrobotrys oligospora, Appl. Environ. Microbiol.,* 47, 1358, 1984.

32. **Rosenzweig, W. D., Premachandran, D., and Pramer, D.,** Role of trap lectins in the specificity of nematode capture by fungi, *Can. J. Microbiol.,* 31, 693, 1985.

33. **Nordbring-Hertz, B., Friman, E., and Mattiasson, B.,** A recognition mechanism in the adhesion of nematodes to nematode-trapping fungi, in *Lectins, Biology, Biochemistry and Clinical Biochemistry,* Vol. 2, Bøg-Hansen, T. C., Ed., W. de Gruyter, Berlin, 1982, 83.

34. **Friman, E., Olsson, S., and Nordbring-Hertz, B.,** Heavy trap formation by *Arthrobotrys oligospora* in liquid culture, *FEMS Microbiol. Ecol.,* 31, 17, 1985.

35. **Borrebaeck, C. A. K., Mattiasson, B., and Nordbring-Hertz, B.,** A fungal lectin and its apparent receptors on a nematode surface, *FEMS Microbiol. Lett.,* 27, 35, 1985.

36. **Borrebaeck, C. A. K., Mattiasson, B., and Nordbring-Hertz, B.,** unpublished data, 1987.

37. **Veenhuis, M., Harder, W., and Nordbring-Hertz, B.,** unpublished data, 1987.

38. **Nordbring-Hertz, B. and Chet, I.,** Fungal lectins and agglutinins, in *Microbial Lectins and Agglutinins: Properties and Biological Activity,* Mirelman, D., Ed., John Wiley & Sons, New York, 1986, 393.

39. **Heintz, C. E. and Pramer, D.,** Ultrastructure of nematode-trapping fungi, *J. Bacteriol.,* 110, 1163, 1972.

40. **Dowsett, J. A. and Reid, J.,** Light microscope observations on the trapping of nematodes by *Dactylaria candida, Can. J. Bot.,* 55, 2956, 1977.

41. **Dowsett, J. A. and Reid, J.,** Observations on the trapping of nematodes by *Dactylaria scaphoides* using optical, transmission and scanning-electron-microscopic techniques, *Mycologia,* 71, 379, 1979.

42. **Dowsett, J. A., Reid, J., and van Caeseele, L.,** Transmission and scanning electron microscope observations on the trapping of nematodes by *Dactylaria brochopaga, Can. J. Bot.,* 55, 2945, 1977.

43. **Veenhuis, M., Nordbring-Hertz, B., and Harder, W.,** Occurrence, characterization and development of different types of microbodies in the nematophagous fungus *Arthrobotrys oligospora, FEMS Microbiol. Lett.,* 24, 31, 1984.

44. **Wimble, D. B. and Young, T. W. K.,** Capture of nematodes by adhesive knobs in *Dactylella lysipaga, Microbios,* 36, 33, 1983.

45. **Tzean, S. S. and Estey, R. H.** Transmission electron microscopy of fungal nematode-trapping devices, *Can. J. Plant Sci.,* 59, 785, 1979.

46. **Veenhuis, M., Nordbring-Hertz, B., and Harder, W.,** Development and fate of electron dense microbodies in trap cells of the nematophagous fungus *Arthrobotrys oligospora, Antonie van Leeuwenhoek,* 51, 399, 1985.

47. **Nordbring-Hertz, B., Zunke, U., Wyss, U., and Veenhuis, M.,** Trap formation and capture of nematodes by *Arthrobotrys oligospora,* Film (C 1622), produced by Inst. Wiss. Film, Göttingen, Germany, 1986.

48. **Olthof, Th. H. A. and Estey, R. H.** A nematotoxin produced by the nematophagous fungus *Arthrobotrys oligospora* Fresenius, *Nature (London),* 197, 514, 1963.

49. **Balan, J., Križková, L., Nemec, P., and Vollek, V.,** Production of nematode-attracting and nematicidal substances by predacious fungi, *Folia Microbiol.,* 19, 512, 1974.

50. **Mankau, R.,** Nematicidal activity of *Aspergillus niger* culture filtrates, *Phytopathology,* 59, 1170, 1969.

51. **Križková, L., Dobias, J., Pódová, M., and Nemec, P.,** Nematocide effect of entomophilic and entomogenous fungi, *Folia Microbiol.,* 24, 171, 1979.

52. **Giuma, A. Y. and Cooke, R. C.,** Nematotoxin production by *Nematoctonus haptocladus* and *N. concurrens, Trans. Br. Mycol. Soc.,* 56, 89, 1971.

53. **Kennedy, N. and Tampion, J.** A nematotoxin from *Nematoctonus robustus, Trans. Br. Mycol. Soc.,* 70, 140, 1978.

54. **Giuma, A. Y., Hackett, A. M., and Cooke, R. C.,** Thermostable nematotoxins produced by germinating conidia of some endozoic fungi, *Trans. Br. Mycol. Soc.,* 60, 49, 1973.

55. **Križková, L., Balan, J., Nemec, P., and Kolozsváry, A.,** Predacious fungi *Dactylaria pyriformis* and *Dactylaria thaumasia:* production of attractants and nematicides, *Folia Microbiol.,* 21, 493, 1976.

56. **Veenhuis, M.,** personal communication, 1987.

57. **Thorn, R. G. and Barron, G. L.,** Carnivorous mushrooms, *Science,* 224, 76, 1984.

58. **Tzean, S. S. and Estey, R. H.,** Nematode-trapping fungi as mycopathogens, *Phytopathology,* 68, 1266, 1978.

59. **Persson, Y., Veenhuis, M., and Nordbring-Hertz, B.,** Morphogenesis and significance of hyphal coiling by nematode-trapping fungi in mycoparasitic relationships, *FEMS Microbiol. Ecol.,* 31, 283, 1985.

Chapter 4

INVERTEBRATE PREDATORS

Richard W. Small

TABLE OF CONTENTS

I. INTRODUCTION

Predators of nematodes continually excite interest amongst nematologists when observed to attack and consume nematode pests. Unfortunately, few nematologists are employed to study such organisms and knowledge of these predators has accumulated slowly. Information is also derived from the work of other biologists concerned with the feeding ecology of the taxa in which predators of nematodes are found. These groups include Protozoa, Turbellaria, Tardigrada, Oligochaeta, Insecta, and Acari as well as Nematoda. The last group has probably attracted most attention, possibly through the double-interest for nematologists of one nematode feeding upon another.

The studies that have been made have been justified on two main grounds; the first is the general contribution to biology and to ecology in particular. The second is that predators of nematodes may have an economic value in reducing the number of nematode pests. The natural, background level of control through predation that must operate on all nematode pests is largely unseen — we are much more aware of the damage wreaked by the survivors. A more exciting, but much less likely possibility is the identification of a predator that could act as a "magic-bullet" in the control of one or more nematode pests in the manner of some of the classic examples of biological control of insects and weeds. These points are discussed further below, but it must be said that none of the predators so far identified has all the requisite qualities of an effective biological control agent; on the other hand there appears to have been no systematic search for such a predator and certainly none in the center of origin of a nematode pest. Our knowledge of the biology of predators of nematodes is so incomplete that we still have little idea of their contribution to natural control. With the increased interest in, and need for, integrated control methods for plant parasitic nematodes,[1] study of their predators is increasingly justified.

II. PROTOZOA

Ciliate protozoans have occasionally been reported to attack nematodes but do not appear to be important predators. *Stylonichia pustulata* Ehrbg. ingested only moribund nematodes[2] and nematodes ingested by *Urostyla* sp. subsequently escaped by rupturing the pellicle of the ciliate.[3] The nematodes suffered no obvious ill-effects and *Urostyla* should not be considered a predator of nematodes. Esser and Sobers also reported ciliates and amoebae that captured nematodes which later escaped or were released.[4]

More interest has been shown in *Theratromyxa weberi* Zwillenberg, an amoeboid protozoan first reported by Weber, Zwillenberg and van der Laan.[5,6] Similar organisms have been described attacking *Meloidogyne incognita* (Kofoid and White), *Globodera rostochiensis* (Woll)., and *Heterodera schachtii* Schmidt.[7,8] In the accounts given by Weber, Zwillenberg and van der Laan[5] and by van der Laan,[9] a creeping trophic form emerged from hypnocysts attached to the cysts of *G. rostochiensis* and often entered the cysts as the juvenile nematodes hatched. Nematodes adhered to the organism on contact and were engulfed in 20 to 120 min. During this period the prey would become motionless and additional juveniles might be caught. Eventually a cyst wall was secreted and the prey was digested over several days; such cysts could contain up to 128 nematodes.[9] A large (300 μm) amoeboid organism emerged from the digestive cyst and in the absence of prey would anastomose with other individuals to form a network up to 30 mm in diameter. This would in time separate into smaller, creeping forms similar to the trophic stage.

Nematodes attacked by *T. weberi* included *Meloidogyne* sp., *Ditylenchus dipsaci* (Kuhn), *Hemicycliophora* sp., *Pratylenchus pratensis* (de Man), and *Rhabditis* sp. Bacteria, algae, and large nematodes such as *Mononchus* and *Dorylaimus* were not attacked. In culture, 8000 juvenile *Pratylenchus* were destroyed by 400 amoebae in 5 days, but in a pot experiment no *T. weberi* were recovered after 6 months.[9] Van der Laan concluded that the slow rate of spread, susceptibility to desiccation, and nonspecific predation limited the biological control value of *T. weberi*, but Paramonov considered that it might control *M. incognita*.[7] Sayre[10] conducted further studies on *T. weberi* in which he showed that juveniles of *Heterodera trifolii* Goffart and adults and juveniles of *Aphelenchus avenae* Bastian, *Aphelenchoides rutgersi* Hooper and Myers, and *Rotylenchus reniformis* Linford and Oliveira as well as juvenile *M. incognita* were utilized as prey. The life cycle was completed in 23 hr at 21°C; details of the life cycle were generally similar to those previously described. In sand cultures, *M. incognita* juveniles were reduced by 79% compared to controls, but in pot experiments with tomatoes, root galling was not reduced. Cysts remained viable for 8 months provided that moisture was available.

T. weberi has been recovered from sandy soils in the Netherlands, a fen soil in England, and a marl soil in Canada. It remains a fascinating organism and one that may contribute to natural control in those soils in which it is found. If that proves to be the case it might be a useful species to introduce to other soils, but it is unlikely to effect an immediate and sustained reduction in phytoparasitic nematode populations to subeconomic levels. Intolerance of desiccation is likely to be the greatest restraint on the selection of sites for such introductions.

III. TURBELLARIA

The only systematic study of soil Turbellaria as predators of nematodes is the work of Sayre and Powers.[11,12] They isolated the rhabdocoel *Adenoplea* sp. from greenhouse soil where it had fed upon *Meloidogyne incognita*. In feeding trials other nematodes consumed were *Panagrellus redivivus* (L.), *Ditylenchus myceliophagus* Goodey, *D. dipsaci*, *Turbatrix aceti* (Muller), *Heterodera trifolii*, and saprophagous nematodes. The feeding rate was estimated as 3.5 *M. incognita* juveniles per hour per adult predator, but particle size of the medium was important in determining the rate of predator-prey encounters. Nematodes varied in their nutritive value; in preliminary trials using constant densities or biomass of prey more eggs were deposited with *P. redivivus* than with *D. myceliophagus* or *M. incognita*. Further experiments showed the following order of prey suitability: *P. redivivus* > *D. dipsaci* > *H. trifolii* > *M. incognita*. Hatching took 9 days but poor hatching rates were obtained when *D. myceliophagus*, *M. incognita*, or *H. trifolii* were used as prey.

In small vials containing soil, predation by *Adenoplea* sp. reduced root galling when tomato plants were planted 48 hr after the addition of 30 turbellarians and 350 *M. incognita*. If the tomatoes were planted at the time of predator and prey inoculation no significant reduction in galling was evident. In discussing *Adenoplea* sp. as a potential control agent Sayre and Powers[11] make three specific points: (1) that adults in natural populations are confined to the soil surface, (2) that the "natural" population on the greenhouse bench had caused no noticeable reduction in the *M. incognita* populations over 6 years, and (3) that densities used to obtain the partial control of *M. incognita* were approximately three times greater than the density of the greenhouse population. In general, known soil Turbellaria carry the advantages and disadvantages of nonspecific predators (shared with many other predators of nematodes), but have the major disadvantage of restriction to the upper layers of moist soils which would preclude their widespread use in biological control programs.

IV. TARDIGRADA

Reports of tardigrades attacking nematodes are scattered throughout the literature, but few species have been subjected to more specific studies.[13-15] Hutchinson and Streu[16] recorded an unidentified tardigrade which attacked and immobilized *Trichodorus aequalis* Allen and

Tylenchus sp. amongst other nematodes. *Macrobiotus richtersii* Murray used the oral stylet to penetrate the cuticle of nematodes and may have secreted a toxin to immobilize the prey which included large species such as mononchids.[3,17] Tardigrades are occasionally numerous in citrus orchards and *Macrobiotus* spp. are reported to feed upon *Tylenchulus semipenetrans* Cobb[18,19] and *Meloidogyne* sp.[20]

Sayre[21] maintained a culture of a tardigrade provisionally identified as *Hypsibius myrops* du Bois-Reymond Marcus on a medium of sterilized *Sphagnum* using *P. redivivus* as prey; the tardigrade held the nematode with its mouthparts and withdrew the body contents through the oral stylet. *M. incognita* and *D. dipsaci* were also eaten. The tardigrade frequently punctured nematodes without feeding; punctured prey were subject to invasion by protozoa and bacteria. Individual tardigrades deposited up to 18 eggs and populations increased from 50 to 5900 over 3 months when well supplied with food.

Tardigrades seem unlikely to be utilized as biological control agents; they are generally associated with habitats that are continuously or periodically moist (e.g., moss) and are not common in mineral soils. In addition predation is nonspecific,[15] although this may allow persistance in the temporary absence of a given prey species. The renowned cryptobiotic abilities of tardigrades would be an advantage and could provide a suitable form for storage and dissemination if required.

V. OLIGOCHAETA

Enchytraeids have received remarkably little attention as potential predators of nematodes. Doncaster[17] recorded *Fridericia* sp. as feeding upon live *Heterodera trifolii* females. Schaerffenberg[22] claimed that individuals of *Fridericia* sp. and *Enchytraeus* sp. entered the roots of sugar beet to feed upon juvenile *H. schachtii* so allowing the plants to recover from the infestation; in pot experiments populations of *H. schachtii* were controlled.[23] Boosalis and Mankau[18] cast doubt on these observations, pointing out that enchytraeids have no means of penetrating healthy roots and are normally considered to feed upon decomposing matter, whereas *H. schachtii* requires sound roots for development. The observation that *Diplenteron colobocercus* Andrassy moved away from enchytraeids (*Hemifridicia* sp.) may be relevant; the strongly alkaline salivary gland secretion of enchytraeids may be nematocidal, although this does not prevent some nematodes from predating upon enchytraeids.[24,25] Further studies are needed to clarify the interactions between enchytraeids and nematodes.

The influence of earthworms (Lumbricidae) on nematode populations has received rather more attention in recent years. Nematodes contributed to the diet of the tropical earthworm *Lampito mauritii* (Kinberg)[26] and active nematodes have been found in the pharyngeal and oesophageal regions, but not in the crop, gizzard, or feces of *Lumbricus terrestris* Linn., suggesting that nematodes contributed to the nutrition of the worms.[27] Unfortunately *L. mauritii* favored predatory nematodes with plant parasitic nematodes being least predated.

Rössner has studied the effect on nematode populations of several earthworm species.[28-30] In greenhouse trials *Eisenia foetida* Savigny ingested *Pratylenchus penetrans* (Cobb) and other *Pratylenchus* spp. in the decaying roots of wheat, alfalfa, and chickweed, leading to a significant reduction in the population of plant parasitic nematodes. Separate trials have also shown population reductions when nematode infested plant tissue is added to soil containing *E. foetida*, including *Pratylenchus* in alfalfa, *Aphelenchoides ritzemabosi* (Schwartz) in chrysanthemum leaves, and *Meloidogyne* in tomato roots. In each case few nematodes were recoverable from pots containing earthworms. Other earthworm species have been shown to reduce nematode populations in similar experiments; *L. terrestris*, *L. rubellus* Hoffmeister, and *Allolobophora caliginosa* Savigny may be of greater importance in that they are more common in mineral soils and prefer cooler conditions than *E. foetida*.[29,31] Rössner's studies have concentrated upon nematodes in plant tissues as he considers that earthworms are unlikely to significantly reduce populations of nematodes dispersed throughout the soil.

Yeates investigated the influence on nematode populations of earthworms inoculated into the soil of New Zealand pastures to improve soil conditions.[32] Marked reductions in total nematode populations (37 and 49% depending on soil type) in plots with earthworms were noted. Generic diversity was unaffected but the genera varied in their response to the presence of earthworms, leading to changes in their contribution to the community. A pot experiment with a third soil type confirmed these results with an even greater decrease in the total nematode numbers (66%). Yeates considers that nematodes contribute to the earthworms' diet but that the environmental modification resulting from earthworm introduction is the main cause of the observed reduction in nematode numbers and changes in community structure.[32] One further particular result may be mentioned as illustrating the interaction between predators of nematodes: at one site the number of predatory mononchid nematodes decreased from $244 \times 10^3/m^2$ to $89 \times 10^3/m^2$ in the presence of earthworms.

VI. INSECTA

Considering the numerous species of insects that spend at least part of their life cycle in soil remarkably few have been recorded as feeding upon nematodes. Most reports, and the only detailed studies, concern Collembola; other records include a dipteran larva feeding upon the *Belonolaimus* sp.[33] and an ant species that prepared food balls composed mainly of fungus, but including nematode eggs and remains.[4] Larvae of the staphylinid beetles *Philanthus* sp., *Trechus* sp., and *Omalium* sp. fed upon live *Heterodera* females and sometimes on cysts.[17]

Amongst the reports of nematode predation by Collembola, Brown describes an *Isotoma* sp. that consumed nematodes in 2 to 3 sec, the prey being eaten from one end.[34] Large numbers of Collembola were associated with organic amendments and with exposed females of *T. semipenetrans*, and some species were maintained on agar for 6 months with nematodes as prey.[18] *Proisotoma minuta* Tullberg may also predate upon *Tylenchulus semipenetrans*.[19] Blackith investigated the influence of collembolan predation by inoculating blocks of peat with *Acrobeloides nanus* (de Man), *Plectus* spp., and the collembolans *Tomocerus* spp., *Lepidocyrtus* spp., and *Folsomia* spp.[35] At the end of the experiment nematode populations were generally greater in the presence of Collembola than in their absence, despite laboratory feeding trials in which *Folsomia* sp. and *Tomocerus longicornis* (Mueller) had eaten *A. nanus* and *Proteroplectus acuminatus* (Bastian). It was not clear why the presence of Collembola should increase nematode populations but an experiment suggested the increased frass was not involved. Nematodes in peat would be able to obtain refuge from Collembola which would at least allow their persistence.

Murphy and Doncaster[36] investigated nematode predation by seven species of Collembola; *Onychiurus armatus* Tullberg was the most voracious and readily attacked the cysts of *Heterodera cruciferae* Franklin, taking 6 to 12 hr to abrade the cyst wall. Once penetrated 4 to 5 predators consumed the cyst contents. *O. armatus* also consumed the white females of *H. cruciferae*, fourth stage juveniles and young females of *H. trifolii*, a mononchid, and dorylaim. Vermiform nematodes were bitten through transversely with no attempt to consume the ends of the prey. In pot experiments *O. armatus* damaged 6.9% of *H. cruciferae* cysts and total predation may have been much greater as white females were poorly recovered by the extraction technique. In field plots of rape up to 30% of *H. cruciferae* cysts were damaged by Collembola, principally *O. armatus*.

In an interesting study that provides a rare insight into the effect of combinations of predators, Sharma[37] utilized two collembolans, *Tullbergia krausbaueri* Borner and *O. armatus* separately and together in a pot experiment on predation of *Tylenchorhynchus dubius*

Table 1
RESULTS OF A POT EXPERIMENT
INVESTIGATING THE INFLUENCE OF
PREDATION ON *TYLENCHORHYNCHUS DUBIUS*
POPULATIONS MAINTAINED ON RYEGRASS

	Final populations of *T. dubius*[a]	
Treatments	**2nd Stage juveniles only**	**All stages**
1. No nematodes, no predators	0	0
2. Nematodes inoculated at same time as predators, no predators[b]	97	508
3. Nematodes only	2065	5567
4. All mite species, both Collembola species	443	2215
5. All mite species, no Collembola species	335	1618
6. *Rhodacarus roseus* only	843	5115
7. *Pergamasus runcatellus* only	1152	4774
8. *Hypoaspis aculeifer* only	250	1785
9. Both Collembola species	764	2702
10. *Tullbergia krausbaueri* only	755	3633
11. *Onychiurus armatus* only	366	2589

[a]   Initial population of *T. dubius* = 25 adults (Treatments 2 to 11); initial inocula of predators = 6 (Treatments 3 to 11), 10 (Treatment 2), 0 (Treatments 1 to 3); 6 replicates of each treatment.

[b]   Treatment 2 terminated 51 days after inoculation of nematodes to indicate population density at time of predator inoculation; experiment terminated 77 days after predator inoculation.

After Sharma, R. D., *Meded. Landbouww. Wageningen*, 71, 1, 1971.

(Butschli) infesting ryegrass (Table 1). Although populations of *T. dubius* were considerably reduced by both species, they were less effective than some of the mite species used (see below). In combination they were amongst the most effective treatments, but much of the predation appears to be attributable to *O. armatus* which achieved a reduction of 50% when used alone. The effect of the predation is particularly noticeable in the reduction of second stage juveniles. The combination of both collembolans and all four mite species did not yield as large a population decrease as did some other treatments, illustrating the complexity of the interactions between predators of nematodes.

Gilmore[38] showed that ten species of Collembola fed readily on nematodes. The fecundity of *Sinella caeca* (Schott) was similar when either nematodes or yeast were supplied as food, but growth rate was greater with nematodes. Nematodes were rapidly digested, casting doubt on the value of gut content analysis as a means of prey determination. Some Collembola were voracious feeders, and could successfully attack heavily cuticularized nematodes such as *Criconemoides* sp. *Entomobryoides dissimilis* (Moniez) consumed over 2000 *Panagrellus* sp. per day in Petri dishes, but the addition of vermiculite provided a refuge for the nematodes that greatly decreased the predation rate. This is one of the few clear demonstrations of a problem that is likely to be encountered in any attempt to utilize microarthropods as control agents — their inability to follow nematodes through the smallest pore spaces.[39] Nematode predation by Collembola undoubtedly contributes to natural control of plant parasitic species and their potential is discussed further below.

VII. ACARI

The literature concerning the predation of nematodes by mites has expanded considerably in the last 15 years and it has become increasingly clear that many mites can utilize nematodes as at least part of their diet. Much of this work has come from acarologists concerned with mite ecology, with the result that the range of prey attacked and the economic importance of mite predation on nematodes has received relatively little attention. This is likely to change as the diversity of mite predators and their nematode prey becomes more widely appreciated.

Nematodes featured prominently in the diet of mites in 4 Gamasid families[40] and 6 predatory mites were amongst the 52 enemies of the *Meloidogyne* spp. recorded by Linford and Oliveira.[20] Isolated reports of nematophagous mites have continued since such early records: *Pergamasus crassipes* L. feeding on live *Heterodera* females and on cysts;[17] *Gaeolaelaps aculeifer* Canestrini feeding on *Xiphinema index* Thorne and Allen;[41] two unidentified mites predating upon *T. semipenetrans*,[19] and the *Hypoaspis* sp. attacking *Meloidogyne* sp. and *Heterodera* sp.[42] Surprisingly few experimental studies were stimulated by these reports but interesting results were obtained from those that were undertaken.

Sharma's investigation of the influence of predatory mites in his study of *T. dubius* biology has already been mentioned (see Table 1). The three mite species in combination appeared to be the most effective in terms of the reduction in *T. dubius* numbers achieved, but *Hypoaspis aculeifer* Can. alone was almost as effective.[37] The presence of Collembola reduced the efficacy of the three mites and illustrates the need for studies that utilize a variety of predators from differing taxonomic groups in order to determine interactions between members of the soil community and hence simulate the soil ecosystem.

In vitro studies with *Lasioceius penicilliger* Berlese showed that the mite captured *T. dubius* easily and devoured them in 5 min although feeding was followed by an unspecified refractory period.[37] Ten mites consumed 90 *T. dubius* in 48 hr. In a pot experiment a 44% reduction in the final *T. dubius* population compared to the mite-free control was achieved; this was sufficient to keep the nematode population constant. In a separate experiment the influence of *Rhodacarus roseus* Oudemans on *T. dubius* populations on ryegrass was studied; after 1 month *T. dubius* numbers in experimental pots were less than one-sixth those in the mite-free controls. Similar single predator studies were conducted by Van de Bund.[43]

Cayrol[44] noted that *Lasioeseius thermophilus* Wilmann, a mite found in mushroom compost and previously believed to feed on mushroom mycelium, consumed the eggs of *D. myceliophagus*. Experiments showed that in the absence of nematodes mite populations declined, but in their presence mite numbers increased from inocula of 10 and 50 to 603 and 1611, respectively. In the absence of mites the nematode population increased from 30 to 504,000 compared to 184,000 (10 mites inoculated) and 18,550 (50 mites inoculated). These results suggest that far from feeding on mushroom mycelium the mite should be encouraged as an effective enemy of *D. myceliophagus*.

The bulb mite *Rhizoglyphus echinopus* (Fumouze and Robin) feeds upon a number of tylenchid and dorylaim phytoparasites but also on the mycelium of *Fusarium* sp., tobacco leaves, and enchytraeids.[45] Small nematodes were completely devoured, larger ones bitten into pieces and the contents withdrawn. Eggs and white females of *Heterodera* spp. were ingested and brown cysts were attacked, although less readily.

A greater stimulus to studies on nematophagous mites was the discovery that two known predators of house-fly eggs, *Macrocheles muscaedomesticae* (Scopoli) and *Glypotholaspis confusa* Foa, also fed upon nematodes.[46] The former fed upon *Rhabditella leptura* (Cobb), *Diplogaster* sp., and *Panagrolaimus* sp. Adult mites preferred house-fly eggs and reproduction was reduced in their absence; reproduction was greatest when both fly eggs and nematodes were available. Proto- and deuto-nymphs preferred nematodes as prey. A self-sustaining food chain could be established using fermented fly larvae rearing medium pre-inoculated with nematodes, but the yield was less than half that obtained under optimum conditions. All motile stages of *G. confusa* also fed upon nematodes. Since this study a number of other workers have used *R. leptura* or other rhabditid nematodes as prey for predatory mesostigmatid mites.[47-51] A rather more detailed study used four manure inhabiting mites and five species of prey;[52] individuals of *M. muscaedomesticae* and *Parasitus* sp. aggregated at the nematodes rapidly and both adults and nymphs fed on all the offered species, whereas few *G. confusa* were observed to feed on nematodes. *Fuscuropoda* sp. aggregated slowly around the nematodes, but remained in the vicinity and fed upon all species.

In 1966 Rockett and Woodring[53] reported that the oribatid mite *Pergalumna omniphagus* Rockett and Woodring was a facultative predator on nematodes such as *Pelodera lambdiensis* (Maupas) and *Tylenchorhynchus martini* Fielding. Although the predation rate was low this report was of considerable interest in that oribatids had been considered nonpredatory. Subsequently, several studies have attempted to establish which mites are predators of nematodes; rather fewer have examined the prey range or aspects of predation such as capture rate or searching behavior.

Muraoka and Ishibashi[54] used *Cephalobus* sp. as potential prey to identify predatory mites in a mixed species assemblage; those feeding on *Cephalobus* were isolated and offered various free-living and phytoparasitic nematodes. Mites were classified as "always" (23 species), "frequently" (12 species), or "occasionally" (5 species) feeding upon nematodes. Of the nematophagous species 17 were mesostigmatid, 22 cryptostigmatid, and 1 astigmatid. This last (*Caloglyphus* sp.) was used in a pot experiment in which the nematode population was initially depressed in the presence of mites but later exceeded that in the mite-free controls. Why nematode numbers should be stimulated by the presence of the mite is as yet unexplained, but a similar phenomenon was noted by Blackith.[35]

The predatory activity of 17 oribatid species was assessed by Rockett using *Chiloplacus* sp. or *Rhabditis* sp. as prey.[55] Three of four *Pergalumna* species readily exhibited nematophagy; *Galumna* sp., *Fuscoceles* sp., and *Nothrus* sp. fed infrequently on nematodes, and *Ceratozetes* sp. chewed but did not devour them. No other predators were noted.

Eight nematode species attacked by *Lasioseius scapulatus* Kennett included second stage juveniles, but not adults or eggs of *M. incognita*.[56] Cultures were established on a potato-dextrose agar/*Rhizoctonia solani* Kuhn/*A. avenae* system. Reproduction was asexual and the life cycle was completed in 8 to 10 days at room temperature; all motile stages were predatory, using the outstretched anterior pair of legs to detect prey. Up to 80 offspring per mite were recorded in population growth experiments in which nematode populations were depressed by up to 99%.

From these accounts the number of known nematophagous mites has increased considerably; in a recent review of nematophagous mites in the cohort Gamasina, Karg[57] lists 33 species in 21 genera and 7 families. Some of these families have panglobal distributions and the abundance of nematophagous mites in soil (37,500 to 370,000/m² in the top 15 cm depending on soil type) suggests that they must have considerable impact on soil nematode populations, at least under some conditions. This is discussed further below, but two communities warrant particular mention.

Mite-nematode interactions in the galleries of wood-boring insects have recently attracted attention;[58-64] three of these may be of considerable importance to the etiology of tree diseases and their control. The first concerns three species of nematophagous mites associated with the pine sawyer beetle *Monochamus alternatus* Hope;[58,59] they may feed upon *Bursaphelenchus xylophilus* (Steiner and Buhrer) a nematode also found in the galleries and which is involved in pine wilt disease. The second involves other mites which are nematophagous as nymphs but which feed upon bark beetles (*Ips* spp.) as adults. Nematodes may therefore be beneficial in enhancing the natural control of the beetles through increased mite populations. Finally, several species of mite have been shown to be predators of *Contortylenchus* spp. which parasitize wood-boring insects. *Dendrolaelaps neodisetus* (Hurlbutt) reduced the incidence of parasitism by *C. brevicomi* (Massey) in the beetle *Dendroctonus frontalis* Zimm. Such nematophagy may hamper attempts to control wood-boring insects using entomogenous nematodes as well as influencing natural control.

The mite-nematode interaction has been shown to be important in the breakdown of litter in arid ecosystems.[65-67] Unlike temperate soils Collembola and oribatid mites are relatively scarce and prostigmatid mites are the dominant microarthropods in such areas. Following initial observations of tydeid mites predating the eggs of bacteriophagous nematodes experiments have indicated that the nematode population is regulated by mite predation. The nematodes in turn graze the bacteria involved in the initial stages of litter breakdown and in the absence of mites the nematodes' grazing is sufficient to retard litter decomposition. Although their effects are likely to be more diffuse in the more diverse microfauna of temperate and humid tropical soils these studies do demonstrate that nematodes and their predators are important in the dynamics of at least some ecosystems.

VIII. NEMATODA

Since Cobb[68] discovered that *Clarkus papillatus* (Bastian) fed upon *T. semipenetrans* there has been considerable discussion of the role of predatory nematodes in natural control and speculation about their potential as biological control agents of phytoparasitic nematodes. With a few notable exceptions, these discussions have not provoked a corresponding interest in, and commitment to, the experimental studies that are needed to end the speculation. Within the last decade, however, there have been a number of studies that have increased our understanding of the biology of predatory nematodes, particularly with regard to their prey.

The early work of Cobb, Steiner and Heinly, Thorne, and Linford and Oliveira has been reviewed previously[4,18,39,69-74] and only a few brief comments need be added. It is interesting to note that these reviews have been almost equally divided between those that consider that research on predatory nematodes is justified by the potential economic benefit and those that do not; it is reassuring that most agree that such work is justified on ecological grounds.

Cobb[75] described several mononchids as predators of nematodes in addition to *C. papillatus;* he also suggested that the ability of some to float might enable their dispersal using irrigation channels, a point also made by Cassidy[76] for *Iotonchus brachylaimus* (Cobb). The "almost inconceivable numbers" of *Mononchus truncatus* Bastian found by Cobb in some slow sand filter beds (e.g., Washington water works where 96% of the nematode fauna were mononchids) led to the suggestion that mononchids might be cultured under similar conditions to supply sufficient numbers to allow their dissemination as control agents.[77] The only recorded attempts at introductions are those of Thorne who, in 1922, transferred *Iotonchus amphigonicus* (Thorne) to cultivated soils in Utah; dispersed colonies persisted for at least 2 years.[78] He also recorded the apparently successful transfer of *C. papillatus*, *Prionchulus muscorum* (Dujardin), and *I. amphigonicus* to date gardens with the aim of controlling *Subanguina radicicola* (Greeff) but the effect, if any, on this target is not recorded.[79]

The first successful culture of a predatory nematode was by Steiner and Heinly[80] using *C. papillatus* (although the egg they depict is reminiscent of *Prionchulus* spp; see Samsoen et al.[81]). Nematodes were kept in water with soil which appeared essential for first stage juvenile development. Various prey were acceptable to the predator and its ability to consume considerable numbers of *S. radicicola*, and by implication other plant parasites, has since elicited much comment; Steiner and Heinly considered that under some circumstances this predator might completely control *S. radicicola*.

A more pessimistic view was reached by Thorne from his studies on mononchids in sugar beet fields infested with *H. schachtii*.[82] After sampling 200 fields for mononchids, 2 were chosen for further study and were sampled irregularly over 2 years. No definite cycles of mononchid numbers were detected, but a population decline occurred sometime between August 1923 and May 1924. On the basis of somewhat circumstantial evidence Thorne suggested that a sporozoan parasite found in a few *M. truncatus* individuals may have been the cause of the decline. Largely on the basis of this hypothesis, but partly also on the apparent quiescence of mononchids during the summer months, Thorne concluded that mononchids were unlikely to be of economic importance. This has since become the most frequently repeated statement concerning mononchids as control agents and, in my view, did much to hinder research on this group of predators.

Certainly the research emphasis subsequently shifted from mononchids to other predatory nematodes. Linford and Oliveira[20,83,84] reported ten species of predatory dorylaims and described their feeding. Unlike mononchids which swallow small prey whole and suck out the body contents of large prey through the wide, armed stoma, (Figure 1), the dorylaims use the odontostyle to penetrate the cuticle of the prey which is rapidly immobilized by the release of the hydrostatic pressure and disruption of the internal tissues. Fluids and soft tissues are withdrawn through the odontostyle. Of greater interest was the feeding of predatory aphelenchs (*Seinura* spp.) which injected a toxic secretion capable of rapidly paralyzing prey much larger than the predator. *Seinura* species have high reproductive rates and very short life cycles under laboratory conditions.[85-88] Cayrol recovered *Seinura oxura* (Paesler) from mushroom compost containing *D. myceliophagus* which had reached a density of 5713/100g compost, suggesting that it was not controlled by *S. oxura*.[44] In experiments *S. oxura* populations reached >70,000 by day 14 and >278,600 by day 21 at 24°C. *D. myceliophagus* populations peaked at 106,000 on day 14 and the subsequent decrease was followed by a catastrophic decline in *S. oxura* numbers. Such population instability is not considered favorable in control agents but the rapid reproduction would be advantageous. Wood lists ten species of nematode predated by *S. demani* (Goodey), but most are free-living species.[89]

Knowledge of predatory diplogasterids, which feed in a similar manner to mononchids, is also based largely on laboratory cultures, e.g., *Mononchoides changi* Goodrich, Hechler, and Taylor and *M. bollingeri* Goodrich, Hechler, and Taylor,[90] *Butlerius micans* Pillai and Taylor,[91] *Fictor anchicoprophaga* (Paramonov), and *Paroigolaimella bernensis* (Steiner);[92] not all of these are known to feed upon nematodes however. In Yeates' study of the predatory behavior of *D. colobocercus* 11 species of nematode prey were identified, but no selection was apparent when 3 species were offered simultaneously.[24] Prey detection was dependent upon contact and predation was proportional to prey density. Male predators tended to be more active and hence have greater predation rates than females; satiation was not demonstrated at the prey densities used. In a study on the predation abilities of *Butlerius degrissei* Grootaert et al. used 20 species of potential prey; most rhabditids were susceptible to predation although adults of *Rhabditis oxycerca* de Man and *Pelodera* sp. frequently escaped.[93] Of the tylenchids used six species were successfully attacked but three others were not; susceptibility appeared to be related to the mode of parasitism: ectoparasites were generally resistant and endoparasites generally susceptible. This differential vulnerability had previously been noted by Esser[33] using mainly dorylaim predators: five phytoparasitic species were vulnerable to attack, whereas many others were resistant. Ectoparasites generally have thicker cuticles than endoparasites and some species may also have a chemical defense.[33,94]

Nelmes developed the numerical assessment of predation rates in his study of *Prionchulus punctatus* Cobb, as well as elucidating other aspects of its biology.[95] His culture technique

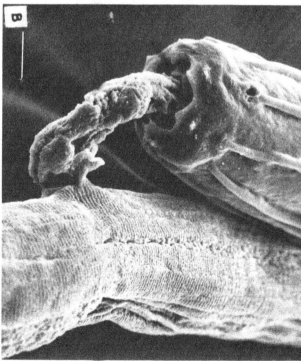

FIGURE 1. Scanning electron micrographs of the monchid predator *Prionchulus punctatus* feeding upon *Panagrellus redivivus*: (A) Ingestion of entire prey, (B) ingestion of body contents. Bars represent 10 μm.

Table 2
SOME RECENT STUDIES ON PREDATORY NEMATODES MAINTAINED IN CULTURE USING SIMILAR METHODS TO THOSE OF NELMES[94]

Predator	Main subject of study	Ref.
Prionchulus punctatus	Feeding behavior, biology	95
	Life cycle, embryology	96
	Culture conditions	97
	Predatory ability	94
	Behavior of males	98
	Population characterization	81
Prionchulus muscorum	Culture conditions, life cycle	99
Clarkus papillatus	Culture conditions, life cycle	99
Mononchus aquaticus	Life cycle, embryology	100
	Description of males	101
	Ultrastructure	102
	Predatory ability	94
Labronema vulvapapillatum Meyl	Feeding mechanism	103
	Life cycle	104
	Predation ability	94
Butlerius degrissei	Life cycle	93
	Predation ability	94
Neoactinolamus duplicidentatus Andrassy	Predation ability	94

proved particularly useful and has enabled further studies on a variety of aspects of predatory nematode biology (Table 2). Small and Grootaert[94] investigated the predation of several predators maintained in such cultures in terms of the predators' ability to attack, wound, and feed upon potential prey. This study indicated that fewer prey may be available to these predators than had previously been thought. Some of the free-living nematodes avoided predation through rapid escape responses (e.g., *Pelodera* sp.), others through thickened or retained cuticles (e.g., *R. oxycerca*). The differing susceptibility of ecto- and endoparasites was confirmed for all predators tested. In this study wounding was considered of equal interest to feeding in that wounded prey were disabled and open to invasion by microorganisms; this would be as effective in control terms as ingestion of prey. In addition, wounding without feeding would not contribute to predator satiation which might otherwise tend to decrease the attack rate. Mononchids were the most efficient predators in this study but the agar medium used does not favor the dorylaim method of attack.

The feeding behavior and prey range of several other predatory nematodes has been investigated recently, e.g., *Mylonchulus dentatus* Jairajpuri,[105] *Mononchus aquaticus* Coetzee,[106] and *Aquatides thornei* Schneider.[107] The predation rate of the last two species remained constant over 10 to 12 days and prey selection or, more likely, variable ease of predation was demonstrated in both species. In these, as in all predatory nematode species so far investigated, no attraction towards potential prey is discernible; labial contact appears essential for prey recognition. Individuals of many predatory nematode species aggregate around prey killed by another individual but not all species show aggregation about an artificially wounded prey. The suggestion has been made that feeding predators may release an attractant.[107] As many predatory nematodes are parthenogenetic and individuals are likely to be surrounded by close relatives in the confines of the soil, kin selection may be invoked to explain this apparent altruism!

Mohandas and Prabhoo extended their laboratory studies on the prey range of six mononchid species from the soils in the Kerala district by analyzing the gut contents of field specimens.[108] Four *Iotonchus* species and *Clarkus mulveyi sensu* Mohandas and Prabhoo ingested prey whole allowing gut content analysis, but *Mylonchulus hawaiiensis* (Cassidy) sucked the body contents from its prey. Of the 19 potential prey tested *Iotonchus kherai* Mohandas and Prabhoo ingested 16, *I. nayari* Mohandas and Prabhoo 14, *C. mulveyi* 9, *I. monhystera* (Cobb) 9, and *I. prabhooi* Mohandas 5. *Macroposthonia ornatum* (Raski), *Caloosia exilis* Mathur, Khan, Nand and Prasad, and *Hoplolaimus mangiferae sensu* Mohandas and Prabhoo were resistant to all six predators. Gut content analysis has also been used by Szczygiel for mononchids in cultivated soils in Poland.[109] Prior to studies such as these information on the diet of predatory nematodes consisted largely of isolated records scattered throughout the literature. These records have recently been compiled and reviewed, with details for 152 named predatory nematodes.[110]

A few studies have taken the investigation of predatory nematodes beyond the laboratory culture. *Thornia* sp. was introduced to pots containing *T. semipenetrans* maintained on citrus.[18] No significant increases in citrus top weight were recorded and the population of *T. semipenetrans* was not significantly reduced in the presence of *Thornia*, but the predator had greatly increased its numbers in the presence of *T. semipenetrans*. Cohn and Mordechai showed that *Mylonchulus sigmaturus* (Cobb) consumed *T. semipenetrans* at the rate of 1.5 to 2.0 prey per hour (twice the rate at which it consumed *Meloidogyne javanica* Treub) under laboratory conditions;[111] a pot experiment in which 50 *M. sigmaturus* were inoculated into naturally *T. semipenetrans* infested citrus orchard soil did not give any significant increase in the weight of sour orange seedlings, but large populations of the predator were consistently associated with small populations of *T. semipenetrans*.[111] Small[112] conducted several pot experiments with various predators; in one a contaminant population of *M. aquaticus* appeared more efficacious than the inoculated *P. punctatus*, but in retrospect the choice of two ectoparasitic species (*Helicotylenchus dihystera* (Cobb) and *Hemicycliophora typica* de Man) as "targets" was unfortunate! Several species of predator used in combination reduced galling on tomato roots caused by *M. incognita* despite a decrease in predator populations.[112]

Ecological studies that include predatory nematodes may aid our assessment of their role in natural control. Generally predatory nematodes contribute <5% to the total nematode population but their contribution to energy flow is greater, partly due to their large size and partly to their active movements.[113-117] In woodland soils, the type of woodland (deciduous or coniferous), the presence or absence of undergrowth, and pH, but not altitude or moisture, were important in determining mononchid numbers.[118] This contrasts with the results of Arpin who demonstrated that mononchids were very susceptible to desiccation;[119] presumably the woodland soils were sufficiently moist at all times. The distribution of predatory nematodes in Massachusetts cranberry bogs,[120] Californian citrus orchards,[19] and cultivated fields in Poland[109] have been recorded. Both Szczygiel[109] and Arpin[121] found that some mononchid species favored particular soils and this was developed by Arpin et al.[122] to suggest that the mononchid community acted as a pedological indicator. The importance of soil conditions is illustrated by the variation in morphometric characters of *P. punctatus* in relation to humus type.[123] Nelmes and McCulloch investigated the population dynamics, depth distribution, and influence of nematocides on mononchids in two sites.[124] In contrast to the results of Thorne,[82] gravid females were always present but the population was equally unstable despite the apparent absence of sporozoans. Bacterial flushes were suggested as an alternative cause of mononchid population declines[124] and this has been linked to their intolerance of high osmotic pressures.[99] Mononchids declined less than the nematode community as a whole when the soil was treated with oxamyl[124] which may be important to integrated control measures.

All the above ecological studies have concerned temperate soils, but recently two studies on mononchids of tropical soils have made a useful contribution. Up to eight species were found in any one site in the Kerala district;[125] the reproductive period of these species varied but no gravid females were recovered in the two driest months. At most sites mononchids constituted 2% of the nematode population but reached 6.7% at one site. In possibly the only recent study to attempt the demonstration of natural control, two sites infested with *H. dihystera* were studied;[126] *I. monhystera* was the dominant predator and gut contents indicated that it predated upon *H. dihystera*. In cultivated soils the peak of predator numbers coincided with that of the prey, but in a pasture peak predator numbers corresponded with the prey minimum. Little can be deduced from a single "cycle", even if that is what these results show, but it may indicate that the relationship between predatory nematodes and their prey is more stable in undisturbed soils.

From the accumulated evidence it is apparent that predatory nematodes may have some limited value as control agents. That not all nematodes are susceptible to these predators weakens the often-voiced criticism that they are polyphagous predators which can exert only partial control at best. It also indicates that the choice of predator and prey may be critical; only the larger mononchids that swallow their prey whole are likely to have any impact on ectoparasitic nematodes, but others may be considered against endoparasites. Particular features, such as the rapid population growth of the *Seinura* spp., need to be borne in mind. The importance of soil type is becoming clearer and there is a need to comprehend the effect of agricultural practices on the predatory nematode community. Much greater understanding of the fundamental biology and ecology of predatory nematodes is essential before we can begin to generalize, draw conclusions, and consider if and how predatory nematodes may be of value.

IX. DISCUSSION AND CONCLUSIONS

Most of the predators of nematodes have been found through chance observations and much of the published information on the predators is descriptive. For Protozoa, Turbellaria, and Tardigrada no new studies have been published in the last 15 years and it would appear that the known forms have no potential as control agents and are probably of ecological significance only in a restricted range of soils and habitats. Insects have fared little better in research terms although the near ubiquitous distribution, relatively high densities and voracity of some species of Collembola[38] suggests that they may contribute significantly to natural control of nematode pests. More studies such as that conducted by Sharma[37] would elucidate their contribution. Earthworms have long been known to be the most important invertebrates in most soils but their role in the population dynamics of nematodes has only recently been recognized. The preference of some species for decaying plant material suggests that they are an important cause of mortality in nematodes causing root necrosis or remaining in shed plant material (e.g., *A. ritzemabosi*).[28-30] The introduction of lumbricid earthworms to New Zealand pastures[32] provides a near-perfect field laboratory and it is to be hoped that studies on earthworm-nematode interactions will continue. If the early indications are confirmed an additional reason, should another be needed, for encouraging and maintaining earthworm populations will have been established.

Mites have received rather more attention in recent years but the emphasis has been on the feeding ecology of the mites and the role of nematodes as an alternative or additional food source to house-fly eggs rather than their role as potential control agents of nematodes. The few studies conducted along this latter line are encouraging but not conclusive.[37,43-45,56] Mites are widespread members of the soil fauna and can occur in high densities;[57] the discovery that some oribatid mites are predatory greatly enlarges the number of known predators of nematodes in many soils. Mites have several advantages in terms of their potential as control agents:[56] the relatively short life cycles and parthogenetic reproduction of many species reduces the time lag in the numerical response to prey increase. In laboratory culture at least, they show the density-dependent predation of an effective and somewhat specialized predator, but it is doubtful that this would apply when the diverse food sources to be found in the natural environment were available. The ability to utilize other food sources may be advantageous in allowing the maintenance of high densities in the absence of the "target". Mites also have greater mobility than some predators of nematodes. Their disadvantages (shared with Collembola) are that they tend to be restricted to, or at least more numerous in the upper levels of the soil, particularly the litter layer, whereas the phytoparasitic nematodes are generally in the rhizosphere. Their generally polyphagous nature reduces their impact on phytoparasitic nematodes and they are unable to follow nematodes through the smallest pore spaces so that the prey have an effective refuge in most soils.[39]

Predatory nematodes also tend to be polyphagous and this is frequently suggested as a contra-indication to their value as control agents. For example, Jones questions the value of detailed studies on predatory nematodes as nonspecific predators rarely exert more than partial control.[39] However, in addition to the advantage of maintaining population numbers in the absence of the target species, polyphagy may be advantageous in that most soils contain more than one plant parasitic nematode species. The selective removal of one may only allow another to take its place and in situations where crop rotations are practiced the nematode causing the greatest damage is likely to change with the crop. When Jones expressed his view it was believed that predatory nematodes fed upon all nematodes they contacted, except perhaps the largest dorylaims such as the *Xiphinema* spp. The studies on the prey of predatory nematodes published in the last 10 years have been as important for showing what these nematodes do not, or can not, prey upon as for indicating available prey.[33,93,94,103-108] It is now apparent that the prey of individual species may be rather limited, with a variety of prey "defenses" preventing predation. These studies also serve to indicate that we may need to choose a predatory nematode to "fit" a given target species; for example, the defenses of the *Helicotylenchus* spp. and *Rotylenchus* spp. are proof against many predators and only those that swallow their prey whole such as *Iotonchus* spp. and *Anatonchus* spp. can effectively predate upon these species.

Predatory nematodes are the only predators of nematodes, with the exception of the Protozoa, that can follow plant parasitic nematodes through the smaller pore spaces. They also tend to be more evenly distributed throughout the soil profile than microarthropods although some (e.g., *Prionchulus* spp.) show a preference for the humus layer of undisturbed soils. They do not have the mobility of microarthropods however, and presumably search less effectively as a result. None of the predators of nematodes investigated appear to be able to detect potential prey at a distance: all appear to rely upon chance contact. Predation is therefore likely to be density-dependent. Mononchids tend to increase the frequency of turns and reversals of direction after contact with prey;[127] this has the effect of increasing the probability of remaining in an area of high prey density (provided that the prey has a patchy distribution) and is a well known behavior pattern in insect predators. Similar behavior is to be expected in other predators of nematodes, but their lack of a distance-chemical sense for the detection of prey remains a serious disability.

From our knowledge of the complexity of the soil environment and its biological community, and from the little that is known of the predators of soil nematodes it seems improbable that a "magic-bullet" control agent will be found. The soil is biologically-buffered by its resident fauna (and flora) and any new introduction would have to be highly competitive to survive.[128] On the other hand, the best biological control agents are not found cohabiting with the target,[56] and as Baker and Cook[129] have commented in discussing fungal pathogens "antagonists should be sought in areas where the disease caused by a given pathogen does not occur, has declined or cannot develop despite the presence of a suitable host rather than where the disease occurs." In examples of successful biological control the agent is frequently obtained from the center of origin of the pest; to my knowledge no such search for predators of plant parasitic nematodes has been conducted. This may be due to a lack of focus upon a specific pest on the part of the researchers on predators.

The presence of predators in most soils and the buffering ability of the soil community suggests that the mere addition of more predators is unlikely to be effective even if sufficient numbers could be reared. At best such additions would need to be repeated at regular and frequent intervals which would negate one of the advantages that biological control can have over chemical control. A more sensible approach would be the manipulation of the soil environment to favor predators; at its simplest this may only require the avoidance of harmful practices. The addition of organic matter appears to favor all the predators for which data is available and represents a relatively easy manipulation of the soil environment. Although Jones suggests this to be a temporary palliative and that the liberation of plant nutrients may be the main source of benefit,[39] many other authors concur in stating that addition of organic matter is beneficial.[57,74,130-133] The reduction in the use of pesticides with their attendant costs and environmental risks is the ultimate aim of much research on predators and parasites of nematodes, and if the addition of organic matter helps to realize that aim, so much the better. If integrated control is to be achieved we must understand the effects of pesticides on the soil fauna, including predators. It is encouraging that the few studies concluded suggest that predators may be relatively resistant to nematocides[124,134,135] if not to some insecticides.[136]

Too often in the past predators of nematodes have been dismissed *en bloc,* revealing a simplistic approach to the complex soil community. We need to recognize the differences as well as the similarities between predators within a single taxon; this will involve assessment of individual species which in itself will probably be discouraging. It will also be misleading if we do not then consider the combined effect of these individual contributions; by doing so we may find that the population dynamics of many free-living and plant parasitic nematodes are, or could be, determined by their predators.

REFERENCES

1. **Bird, G. W. and Thomason, I. J.,** Integrated pest management: the role of nematology, *Bioscience,* 30, 670, 1980.
2. **Doncaster, C. C.,** Electronic flash in photomicrography, *Nematologica,* 1, 51, 1956.
3. **Doncaster, C. C. and Hooper, D. J.,** Nematodes attacked by protozoa and tardigrades, *Nematologica,* 6, 333, 1961.
4. **Esser, R. P. and Sobers, E. K.,** Natural enemies of nematodes, *Proc. Soil Crop Sci. Soc. Fla.,* 24, 326, 1964.
5. **Weber, A. P., Zwillenberg, L. O., and van der Laan, P. A.,** A predacious amoeboid organism destroying larvae of the potato root eelworm and other nematodes, *Nature,* 169, 834, 1952.
6. **Zwillenberg, L. O.,** *Theratromyxa weberi,* a new proteomyxean organism from soil, *Antonie van Leeuwenhoek, J. Microbiol. Serol.,* 19, 101, 1953.
7. **Paramonov, A. A.,** An amoeboid organism destroying infective larvae of the root knot nematode, *Tr. Gelmintol. Lab. Akad. Nauk USSR,* 7, 50, 1954.
8. **Winslow, R. D. and Williams, T. D.,** Amoeboid organisms attacking larvae of the potato root eelworm (*Heterodera rostochiensis* Woll.) in England and the beet eelworm (*H. schachtii* Schm.) in Canada, *T. Pl. Ziekten.,* 63, 242, 1957.
9. **van der Laan, P. A.,** Nader onderzoek over het aaltjesvangende amoeboide organisme, *Theratromyxa weberi* Zwillenberg, *T. Pl. Ziekten,* 60, 139, 1954.
10. **Sayre, R. M.,** *Theratromyxa weberi,* an amoeba predatory on plant-parasitic nematodes, *J. Nematol.,* 5, 258, 1973.
11. **Sayre, R. M. and Powers, E. M.,** A predacious soil turbellarian that feeds on free-living and plant-parasitic nematodes, *Nematologica,* 12, 619, 1966.
12. **Sayre, R. M. and Powers, E. M.,** Effects of temperature on the fecundity and hatch of the soil flatworm reared on free-living and plant-parasitic nematodes, *Phytopathology,* 57, 828, 1967.

13. Le Gros, A. E., How to begin to study tardigrades, *Countryside*, 18, 322, 1958.
14. Cayrol, J. C., Relations ecologiques entre Nematodes et autres organismes terricoles, *Bull. Soc. Zool. Fr.*, 101, 872, 1976.
15. Morgan, C. I., Population dynamics of two species of tardigrada, *Macrobiotus hufelandii* (Schultze) and *Echiniscus (Echiniscus) testudo* (Doyere) in roof moss from Swansea, *J. Anim. Ecol.*, 46, 263, 1977.
16. Hutchinson, M. T. and Streu, H. T., Tardigrades attacking nematodes, *Nematologica*, 5, 149, 1960.
17. Doncaster, C. C., Predators of soil nematodes (film), *Parasitology*, 52, 19P, 1962.
18. Boosalis, M. G. and Mankau, R., Parasitism and predation of soil micro-organisms, in *Ecology of Soil-borne Plant Pathogens*, Baker, K. F. and Snyder, W. C., Eds., University of California Press, Berkeley, 1965, 374.
19. Stirling, G. R. and Mankau, R., Biological control of nematode parasites of citrus by natural enemies, *Proc. Int. Soc. Citricult.*, 3, 843, 1977.
20. Linford, M. B. and Oliveira, J. M., Potential agents of biological control of plant parasitic nematodes, *Phytopathology*, 28, 14, 1938.
21. Sayre, R. M., A method for culturing a predaceous tardigrade on the nematode *Panagrellus redivivus*, *Trans. Am. Microsc. Soc.*, 88, 266, 1969.
22. Schaerffenberg, B., Untersuchungen über die Bedeutung der Enchytraeiden als Humusbildner und Nematodenfeinde, *Z. Pflanzenkr. Pflanzen Pathol. Pflanzenschutz.*, 57, 183, 1950.
23. Schaerffenberg, B. and Tendl, H., Untersuchung über das Verhalten der Enchytraeiden gegenüber dem Zuckerrübennematoden *H. schachtii* Schm., *Z. Angew. Entomol.*, 32, 476, 1951.
24. Yeates, G. W., Predation by *Mononchoides potohikus* (Nematoda, Diplogasteridae) in laboratory culture, *Nematologica*, 15, 1, 1969.
25. Yeates, G. W., An analysis of annual variation of the nematode fauna in dune sand, at Himatangi Beach, New Zealand, *Pedobiologia*, 8, 173, 1968.
26. Dash, M. C., Senapati, B. K., and Mishra, C. C., Nematode feeding by tropical earthworms, *Oikos*, 34, 322, 1980.
27. Piearce, T. G. and Phillips, M. J., The fate of ciliates in the earthworm gut: an *in vitro* study, *Microb. Ecol.*, 5, 313, 1980.
28. Rössner, J., Einfluss von Regenwürmern auf phytoparasitäre Nematoden, *Nematologica*, 27, 340, 1981.
29. Rössner, J., Effects of earthworm feeding on numbers of some phytoparasitic nematodes, in Abstr. 14th Symp. Eur. Soc. Nematologists, St. Andrews, Scotland, 1982, 57.
30. Rössner, J., Wechselwirkungen zwischen Nematoden und anderen Bodenorganismen, *Beurteilung von Bodenbearbeitungssystemen Justus-Leibig Universität*, Giessen, W. Germany, 112, 1983.
31. Rössner, J., personal communication, 1985.
32. Yeates, G. W., Soil nematode populations depressed in the presence of earthworms, *Pedobiologia*, 22, 191, 1981.
33. Esser, R. P., Nematode interactions in plates of non-sterile water agar, *Proc. Soil Crop Sci. Soc. Fla.*, 23, 121, 1963.
34. Brown, W. L., Collembola feeding on nematodes, *Ecology*, 35, 421, 1954.
35. Blackith, R. M., Interrelationships between small arthropods and nematodes in peat, *Proc. Roy. Irish Acad.*, 75B, 531, 1975.
36. Murphy, P. W. and Doncaster, C. C., A culture method for soil meiofauna and its application to the study of nematode predators, *Nematologica*, 2, 202, 1957.
37. Sharma, R. D., Studies on the plant parasitic nematode *Tylenchorhynchus dubius*, *Meded. Landbouww. Wageningen*, 71, 1, 1971.
38. Gilmore, S. K., Collembola predation on nematodes, *Search Agric. Geneva*, 1, 1, 1970.
39. Jones, F. G. W., Control of nematode pests, background and outlook for biological control, in *Biology in Pest and Disease Control*, Price-Jones, D., and Solomon, M. E., Eds., Blackwell, Oxford, 1974, 249.
40. Weis-Fogh, T., Ecological investigations on mites and collemboles in the soils, *Nat. Jutl.*, 1, 135, 1948.
41. Kermarrec, A., Parasites et predateurs de nematodes. Contribution a l'etude de certains facteurs intervenant sur la biologie d'hyphomycetes, D.E.A. Sc. Nat. (thesis), University de Montpellier, Montpellier, France, 1969.
42. Inserra, R. N. and Davis, D. W., *Hypoaspis* nr. *aculeifer* a mite predacious on root knot and cyst nematodes, *J. Nematol.*, 15, 324, 1983.
43. Van de Bund, C. F., Some observations on predatory action of mites on nematodes, *Zesz. Probl. Post. Nauk Roln.*, 129, 103, 1972.
44. Cayrol, J. C., Action des autres composants de la biocenose du champignon de couche sur le nematode mycophage, *Ditylenchus myceliophagus* J. B. Goodey, 1958, et etude de son anabiose: forme de survie en conditions defavorables, *Rev. Ecol. Biol. Sol*, 7, 409, 1970.
45. Sturhan, D. and Hampel, G., Pflanzenparasitische Nematoden als Beute der Wurzelmilbe *Rhizoglyphus echinopus* (Acarina, Tyroglyphidae), *Anz. Schadsl. Pflanzen. Umwelt.*, 50, 115, 1977.

46. **Rodriguez, J. G., Wade, C. F., and Wells, C. N.**, Nematodes as natural food for *Macrocheles muscaedomesticae* (Acarina: Macrochelidae), a predator of the housefly egg, *Ann. Ent. Soc. Am.*, 55, 507, 1962.

47. **Rodriguez, J. G.**, Nutritional studies in the Acarina, *Acarologia*, 6, 324, 1964.

48. **Singh, P. and Rodriguez, J. G.**, Food for macrochelid mites (Acarina) by an improved method for mass rearing of a nematode, *Rhabditella leptura*, *Acarologia*, 8, 549, 1966.

49. **Singer, G. and Krantz, G. W.**, The use of nematodes and oligochaetes for rearing predatory mites, *Acarologia*, 9, 485, 1967.

50. **Ito, Y.**, Changes of the population density and stage compositions of three mesostigmatid mite species on a restricted food supply, *Jap. J. Appl. Ent. Zool.*, 21, 74, 1977.

51. **Ito, Y.**, Predatory activity of mesostigmatid mites (Acarina: Mesostigmata) for housefly eggs and larvae under feeding of nematodes, *Jpn. J. Sanit. Zool.*, 28, 167, 1977.

52. **Ito, Y.**, Predation by manure-inhabiting Mesostigmatids (Acarina: Mesostigmata) on some free-living nematodes, *Appl. Ent. Zool.*, 6, 51, 1971.

53. **Rockett, C. L. and Woodring, J. P.**, Oribatid mites as predators of soil nematodes, *Ann. Ent. Soc. Am.*, 59, 669, 1966.

54. **Muraoka, M. and Ishibashi, N.**, Nematode feeding mites and their feeding behavior, *Appl. Ent. Zool.*, 11, 1, 1976.

55. **Rockett, C. L.**, Nematode predation by Oribatid mites (Acari: Oribatida), *Internat. J. Acarol.*, 6, 219, 1980.

56. **Imbriani, I. and Mankau, R.**, Studies on *Lasioseius scapulatus*, a mesostigmatid mite predaceous on nematodes, *J. Nematol.*, 15, 523, 1983.

57. **Karg, W.**, Verbreitung und Bedeutung von Raubmilben der Cohors Gamasina als Antagonisten von Nematoden, *Pedobiologia*, 25, 419, 1983.

58. **Tamura, H. and Enda, N.**, Life histories of three species of nematode-feeding mesostigmatid mites associated with the pine sawyer beetle *Monochamus alternatus*, *J. Jpn. For. Soc.*, 62, 301, 1980.

59. **Tamura, H. and Enda, N.**, Mesostigmatid mites associated with Japanese Pine Sawyer Beetle, *Jpn. J. Appl. Ent. Zool.*, 24, 54, 1980.

60. **Kinn, D. N.**, Notes on the life cycle and habits of *Digamasellus quadrisetus* (Mesostigmata: Digamasellidae), *Ann. Ent. Soc. Am.*, 60, 862, 1967.

61. **Kinn, D. N.**, Mutualism between *Dendrolaelaps neodisetus* and *Dendroctonus frontalis*, *Environ. Entomol.*, 9, 756, 1980.

62. **Kinn, D. N.**, The life cycle of *Proctolaelaps dendroctoni* Lindquist and Hunter (Acari: Ascidae): a mite associated with pine bark beetles, *Int. J. Acarol.*, 9, 205, 1983.

63. **Kinn, D. N.**, Life cycle of *Dendrolaelaps neodisetus* (Mesostigmata: Digamasellidae), a nematophagous mite associated with pine bark beetles (Coleoptera: Scolytidae), *Environ. Entomol.*, 13, 1141, 1984.

64. **Kinn, D. N. and Witcosky, J. J.**, The life cycle and behavior of *Macrocheles boudreauxi* Krantz, *Z. Ang. Ent.*, 84, 136, 1977.

65. **Santos, P. F. and Whitford, W. G.**, The effects of microarthropods on litter decomposition in a Chihuahuan desert ecosystem, *Ecology*, 62, 654, 1981.

66. **Santos, P. F., Phillips, J., and Whitford, W. G.**, The role of mites and nematodes in early stages of buried litter decomposition in a desert, *Ecology*, 62, 664, 1981.

67. **Elkins, N. Z. and Whitford, W. G.**, The role of microarthropods and nematodes in decomposition in a semi-arid ecosystem, *Oecologia Berlin*, 55, 303, 1982.

68. **Cobb, N. A.**, Citrus root nematode, *J. Agric. Res.*, 2, 217, 1914.

69. **Christie, J. R.**, Biological control — predaceous nematodes, in *Nematology: Fundamentals and Recent Advances with Emphasis on Plant Parasitic and Soil Forms*, Sasser, J. N. and Jenkins, W. R., Eds., University of North Carolina Press, Chapel Hill, 1960, chap. 46.

70. **Sayre, R. M.**, Biotic influences in soil environment, in *Plant Parasitic Nematodes*, Vol. 1, Zuckerman, B. M., Mai, W. F., and Rohde, R. A., Eds., Academic Press, New York, 1971, chap. 9.

71. **Webster, J. M.**, Nematodes and biological control, in *Economic Nematology*, Webster, J. M., Ed., Academic Press, London, 1972, chap. 19.

72. **Ritter, M. and Laumond, C.**, Perspectives d'emploi des nematodes dans les programmes de lutte biologique contre les parasites et les ravageurs des plantes cultivees, presented at Semaine d'Etude Agriculture et Hygiene de Plantes, Gembloux, France, September 8 to 12, 1975, 331.

73. **Arpin, P.**, Les elements predateurs de la microfaune du sol, in *Actualities d'Ecologie Forestiere*, Pesson, P., Ed., Gauthier-Villars, Paris, 1980, 507.

74. **Mankau, R.**, Biological control of nematode pests by natural enemies, *Annu. Rev. Phytopathol.*, 18, 415, 1980.

75. **Cobb, N. A.**, The mononchs: a genus of free-living predatory nematodes, *Soil Sci.*, 3, 431, 1917.

76. **Cassidy, G. H.**, Some mononchs of Hawaii, *Hawaii. Plant. Rec.*, 35, 305, 1931.

77. **Cobb, N. A.,** Transference of nematodes (mononchs) from place to place for economic purposes, *Science,* 51, 640, 1920.
78. **Thorne, G.,** Utah nematodes of the genus *Mononchus, Trans. Am. Microscop. Soc.,* 43, 157, 1924.
79. **Thorne, G.,** Apparent successful transfer of *Mononchus papillatus, M. muscorum* and *M. amphigonicus* from Utah to California, *J. Parasitol.,* 19, 90, 1932.
80. **Steiner, G. and Heinly, H.,** The possibility of control of *Heterodera radicicola* and other plant injurious nematodes by means of predatory nematodes, especially *Mononchus papillatus, J. Wash. Acad. Sci.,* 12, 367, 1922.
81. **Samsoen, L., Arpin, P., Khan, S. H., and Coomans, A.,** Differentiation of *Prionchulus muscorum* (Dujardin, 1845) Wu and Hoeppli, 1929 and *Prionchulus punctatus* (Cobb, 1917) Andrassy, 1958 by eggshell structure, *Rev. Nematol.,* 7, 315, 1984.
82. **Thorne, G.,** The life history, habits and economic importance of some mononchs, *J. Agric. Res.,* 34, 265, 1927.
83. **Linford, M. B.,** The feeding of some hollow-stylet nematodes, *Proc. Helminthol. Soc. Wash.,* 4, 41, 1937.
84. **Linford, M. B. and Oliveira, J. M.,** The feeding of hollow-spear nematodes on other nematodes, *Science,* 85, 295, 1937.
85. **Hechler, H. C.,** Description, developmental biology and feeding habits of *Seinura tenuicaudata* (De Man) J. B. Goodey 1960 (Nematoda: Aphelenchoididae), a nematode predator, *Proc. Helminthol. Soc. Wash.,* 30, 183, 1963.
86. **Hechler, H. C. and Taylor, D. P.,** Taxonomy of the genus *Seinura* (Nematoda: Aphelenchoididae) with description of *S. celeris* n. sp. and *S. steineri* n. sp., *Proc. Helminthol. Soc. Wash.,* 32, 205, 1965.
87. **Hechler, H. C. and Taylor, D. P.,** The life histories of *Seinura celeris, S. oliveirae, S. oxura* and *S. steineri* (Nematoda: Aphelenchoididae), *Proc. Helminthol. Soc. Wash.,* 33, 71, 1966.
88. **Hechler, H. C. and Taylor, D. P.,** The moulting process in species of *Seinura* (Nematoda: Aphelenchoididae), *Proc. Helminthol. Soc. Wash.,* 33, 90, 1966.
89. **Wood, F. H.,** Biology of *Seinura demani* (Nematoda: Aphelenchoididae), *Nematologica,* 20, 347, 1974.
90. **Goodrich, M., Hechler, H. C., and Taylor, D. B.,** *Mononchoides changi* n. sp. and *M. bollingeri* n. sp. (Nematoda: Diplogasterinae) from a waste treatment plant, *Nematologica,* 14, 25, 1968.
91. **Pillai, J. K. and Taylor, D.,** *Butlerius micans* n. sp. (Nematoda, Diplogasterinae) from Illinois with observations on its feeding habits and a key to the species of *Butlerius* Goodey 1929, *Nematologica,* 14, 89, 1968.
92. **Pillai, J. K. and Taylor, D. P.,** Biology of *Paraigolaimella bernensis* and *Fictor anchicoprophaga* (Diplogasterinae) in laboratory culture, *Nematologica,* 14, 159, 1968.
93. **Grootaert, P., Jaques, A., and Small, R. W.,** Prey selection in *Butlerius* sp. (Rhabditida, Diplogasteridae), *Med. Fac. Landbouww. Rijksuniv. Gent.,* 42, 1559, 1977.
94. **Small, R. W. and Grootaert, P.,** Observations on the predation abilities of some soil dwelling predatory nematodes, *Nematologica,* 29, 109, 1983.
95. **Nelmes, A. J.,** Evaluation of the feeding behaviour of *Prionchulus punctatus* (Cobb), a nematode predator, *J. Anim. Ecol.,* 43, 553, 1974.
96. **Maertens, D.,** Observations on the life cycle of *Prionchulus punctatus* (Cobb, 1917) and culture conditions, *Biol. Jb. Dodonaea,* 43, 197, 1975.
97. **Small, R. W. and Evans, A. A. F.,** Experiments on the population growth of the predatory nematode *Prionchulus punctatus* in laboratory culture with observations on life history, *Rev. Nematol.,* 4, 261, 1981.
98. **Samsoen, L.,** Experiments with the male of *Prionchulus punctatus* (Cobb, 1917) Andrassy 1958, *Rev. Nematol.,* 7, 417, 1984.
99. **Arpin, P.,** Etude et discussion sur un milieu de culture pour Mononchidae (Nematoda), *Rev. Ecol. Biol. Sol.,* 13, 629, 1976.
100. **Grootaert, P. and Maertens, D.,** Cultivation and life cycle of *Mononchus aquaticus, Nematologica,* 22, 173, 1976.
101. **Small, R. W. and Grootaert, P.,** Description of the male of *Mononchus aquaticus* Coetzee, 1968 (Nematoda: Mononchidae) with observations on the females, *Biol. Jb. Dodonaea,* 45, 162, 1977.
102. **Grootaert, P. and Wyss, U.,** Ultrastructure and function of the anterior feeding apparatus in *Mononchus aquaticus, Nematologica,* 25, 163, 1979.
103. **Wyss, U. and Grootaert, P.,** Feeding mechanisms in *Labronema* sp., *Med. Fac. Landbouww. Rijksuniv. Gent,* 42, 1521, 1977.
104. **Grootaert, P. and Small, R. W.,** Aspects of the biology of *Labronema vulvapapillatum* (Meyl) (Nematoda, Dorylaimidae) in laboratory culture, *Biol. Jb. Dodonaea,* 50, 135, 1982.
105. **Jairajpuri, M. S. and Azmi, M. I.,** Some studies on the predatory behaviour of *Mylonchulus dentatus, Nematol. Medit.,* 6, 205, 1978.
106. **Bilgrami, A. L., Ahmad, I., and Jairajpuri, M. S.,** Observations on the predatory behaviour of *Mononchus aquaticus, Nematol. Medit.,* 12, 41, 1984.

107. **Bilgrami, A. L., Ahmad, I., and Jairajpuri, M. S.,** Predatory behaviour of *Aquatides thornei* (Nygolaimina: Nematoda), *Nematologica*, 20, 457, 1984.

108. **Mohandas, C. and Prabhoo, N. R.,** The feeding behaviour and food preferences of predatory nematodes (Mononchida) from the soils of Kerala (India), *Rev. Ecol. Biol. Sol*, 17, 53, 1980.

109. **Szczygiel, A.,** Wystepowanie drapieznych nicieni z rodziny Mononchidae w glebach uprawnych w Polsce, *Zesze. Probl. Postep. Nauk. Roln.*, 121, 145, 1971.

110. **Small, R. W.,** A review of the prey of predatory soil nematodes, *Pedobiologia*, 30, 179, 1987.

111. **Cohn, E. and Mordechai, M.,** Experiments in suppressing citrus nematode populations by use of a marigold and a predacious nematode, *Nematol. Medit.*, 2, 43, 1974.

112. **Small, R. W.,** The effects of predatory nematodes on populations of plant parasitic nematodes in pots, *Nematologica*, 25, 94, 1979.

113. **Overgaard-Nielsen, C.,** Studies on the soil microfauna. II. The soil inhabiting nematodes, *Natura Jutl.*, 2, 1, 1949.

114. **Yuen, P. H.,** The nematode fauna of the regenerated woodland and grassland of Broadbalk Wilderness, *Nematologica*, 12, 195, 1966.

115. **Wasilewska, L.,** Nematodes of the dunes in the Kampinos forest. II. Community structure based on numbers of individuals, state of biomass and respiratory metabolism, *Ekol. Polska*, 19, 651, 1971.

116. **Yeates, G. W.,** Nematoda of a Danish beech forest. I. Methods and general analysis, *Oikos*, 23, 178, 1972.

117. **Yeates, G. W.,** Nematoda of a Danish beech forest. II. Production estimates, *Oikos*, 24, 179, 1973.

118. **Boag, B.,** Nematodes associated with forest and woodland trees in Scotland, *Ann. Appl. Biol.*, 77, 41, 1974.

119. **Arpin, P.,** Etude preliminaire d'un facteur ecologique important pour les Nematodes: l'humidite actuelle du sol, *Rev. Ecol. Biol. Sol.*, 6, 429, 1969.

120. **Paracer, S. M., Brzeski, M. W., and Zuckerman, B. M.,** Nematophagous fungi and predaceous nematodes associated with cranberry soils in Massachusetts, *Plant Dis. Rep.*, 50, 583, 1966.

121. **Arpin, P.,** Ecologie et systematique des nematodes Mononchides des zones forestieres et herbacees sous climat tempere humide. I. Types de sol et groupes specifiques, *Rev. Nematol.*, 2, 211, 1979.

122. **Arpin, P., Ponge, J-F., Dabin, B., and Mori, A.,** Utilisation des nematodes Mononchida et des Collemboles pour caracteriser des phenomenes pedobiologiques, *Rev. Ecol. Biol. Sol*, 21, 243, 1984.

123. **Arpin, P. and Ponge, J-F.,** Etude des variations morphometriques de *Prionchulus punctatus* (Cobb, 1917) Andrassy, 1958, *Rev. Nematol.*, 7, 315, 1984.

124. **Nelmes, A. J. and McCulloch, J. S.,** Numbers of mononchid nematodes in soils sown to cereals and grasses, *Ann. Appl. Biol.*, 79, 231, 1975.

125. **Mohandas, C. and Prabhoo, N. R.,** Predatory nematodes (Mononchida) in the soil. Their population fluctuation and utility as biological control agents, in *Proc. Symp. Environmental Biology*, Trivandrum, India, 1980, 92.

126. **Azmi, M. I.,** Predatory behaviour of nematodes. I. Biological control of *Helicotylenchus dihystera* through the predacious nematode *Iotonchus monhystera*, *Indian J. Nematol.*, 13, 1, 1983.

127. **Small, R. W.,** unpublished observations, 1974—1979.

128. **Van Gundy, S. D.,** Nonchemical control of nematodes and root-infecting fungi, in *Pest Control Strategies for the Future*, National Academy of Sciences, Washington, D.C., 1972, 317.

129. **Baker, K. F. and Cook, R. J.,** *Biological Control of Plant Pathogens*, W. H. Freeman & Co., San Francisco, Calif., 1974.

130. **Hutchinson, M. T., Reed, J. P., and Pramer, D.,** Observations on the effects of decaying vegetable matter on nematode populations, *Plant Dis. Rep.*, 44, 400, 1960.

131. **Mankau, R.,** Reduction of root-knot disease with organic amendments under semifield conditions, *Plant Dis. Rep.*, 52, 315, 1968.

132. **Miller, P. M.,** Non-chemical control of plant-parasitic nematodes, *Morris Arbor. Bull.*, 22, 70, 1971.

133. **Good, J. M.,** Bionomics and integrated control of plant parasitic nematodes, *J. Environ. Qual.*, 1, 382, 1972.

134. **Edwards, C. A. and Lofty, J. R.,** Nematocides and the soil fauna, in *Proc. 6th Br. Insecticide Fungicide Conference*, Vol. 1, British Crop Protection Council, London, 1971, 158.

135. **Heijbroek, W. and Van de Bund, C. F.,** The influence of some agricultural practices on soil organisms and plant establishment of sugar beet, *Neth. J. Plant Pathol.*, 88, 1, 1982.

136. **Martin, N. A. and Yeates, G. W.,** Effect of four insecticides on the pasture ecosystem, *N. Z. J. Agric. Res.*, 18, 307, 1975.

Chapter 5

BIOLOGICAL CONTROL OF PLANT-PARASITIC NEMATODES

Graham R. Stirling

TABLE OF CONTENTS

I. INTRODUCTION

It is generally accepted that biological control is a broad concept which encompasses a range of control strategies including cultural practices, host plant resistance, and the introduction or encouragement of antagonistic organisms. Since some of these topics are beyond the scope of this book, this discussion will concentrate on the nematode control achieved by natural enemies as a result of parasitism, predation, competition, or antibiosis. Baker and Cook's[1] definition of biological control has been adopted and modified for nematodes as follows: reduction in nematode damage by organisms antagonistic to nematodes through the regulation of nematode populations and/or a reduction in the capacity of nematodes to cause damage, which occurs naturally or is accomplished through the manipulation of the environment or by the mass introduction of antagonists. Such a definition recognizes that reduction in nematode damage is the primary aim of any control measure and that it need not necessarily be accomplished by reducing nematode numbers. It embraces situations, for example, where control is achieved with antagonists which restrict the movement or invasion of nematodes rather than killing them.

Nematologists have been interested in using natural enemies to control nematodes since the 1920s and 1930s, when Cobb[2] showed an interest in the predatory nematodes, and Linford and Yap[3] conducted experiments with the nematode-trapping fungi. Despite a considerable amount of interest and many optimistic pronouncements during the last 50 years, only a few examples of natural biological control of nematodes have been reported and there are no examples of the widespread commercial use of introduced biological control agents. This situation is in marked contrast to fields such as entomology and weed science where there are many well documented examples of biological control.[4] Even in the related field of plant pathology, where interest in biocontrol has quickened during the last decade or so, biological control now offers answers to many serious disease problems in modern agriculture.[5] The relatively recent recognition of the importance of plant-parasitic nematodes, the need to develop basic information on their taxonomy, physiology, biology, and ecology, the availability of cheap, effective nematicides, and the complexity of the soil environment are perhaps some of the reasons why developments in biological control of nematodes have lagged behind those in other fields.

As we approach the end of the twentieth century there has never been a greater need to find alternative methods of nematode control. In some countries, the fumigants dibromochloropropane and ethylene dibromide, which have formed the basis of nematode control programs on many crops since the end of World War II, have been removed from the market because of health and environmental considerations. The use of others such as methyl bromide and 1,3 dichloropropene is being reviewed, while the long-term future of the nonvolatile nematicides, seen as alternatives to the fumigants in many crops, is clouded following the recent discovery of some of them in groundwater. With the increasing cost of testing and registering pesticides, the development of new nematicides has almost ground to a halt, so that additional nonchemical means of nematode control will have to be found. Biological control will play an increasing role in practical nematode control in the future and this chapter aims to evaluate the present position, some of the problems limiting its development, and some of the prospects for the future.

II. POPULATION REGULATION

All plant-parasitic nematodes spend at least some part of their lives in soil, one of the most complex of habitats. Within that environment nematode populations do not increase indefinitely because they are constrained by a number of physical and biological factors. There are many excellent discussions on the dynamics of animal populations and the mechanisms which regulate them[6-10] and this topic will not be discussed in detail here. Suffice to say that two major forces regulate natural populations: those that are not influenced by the density of the organism and those that are density dependent.

The mortality factors that operate independent of density are largely meteorological and physical. In the soil environment they include extremes of moisture and temperature, and normal agricultural practices such as plowing and plant destruction. They also include various polyphagous parasites and predators. Although these antagonists do not depend on a specific organism for food, they may take a relatively constant toll of that organism incidental to their survival on other substrates. In studies with insects, physical hazards, and nonspecific antagonists have been shown to have drastic effects on the number of insects in a population and may be the major factors limiting abundance when insect populations are at their lowest level. However, they do not regulate those populations.[10]

Population regulation occurs through density dependent mechanisms; extreme densities are inevitably met by starvation or lack of space, if not predation, disease, or some other catastrophe associated with such abundance. Although density independent factors have an impact on population size, density dependent forces are the key to the long-term suppression and management of pest populations by biological means. At low pest densities, populations of host-specific parasites and predators are low, increasing the chances of the pest surviving and allowing it to increase its population. As the pest begins to increase, so does the antagonist, but with some lag time. Eventually the parasite or predator overwhelms the host, leading to a drastic decline in pest populations followed by a decline in the parasites and predators due to inadequate food.

The challenge facing those interested in biological control is to develop means of using biological mechanisms of population regulation and/or suppression to reduce the damage caused by plant-parasitic nematodes. Ideally, systems of biological control should be stable, permanent, and self-sustaining, and such systems are only likely to be achieved with antagonists which regulate nematode populations through density-dependent mechanisms. Such antagonists tend to be relatively host-specific and usually are intimately associated with their hosts. Provided they have adequate means of surviving periods of low host density their population regulating mechanisms should enable them to permanently constrain nematode populations within certain limits. Those antagonists able to maintain an equilibrium nematode density at or below the economic threshold have the greatest potential for use as biological control agents in agriculture. However, this does not mean that organisms which suppress nematode populations rather than regulate them do not have a place in biological control. Nematicides are effective only for limited periods and there is no reason why antagonists which act independently of nematode density could not be used in much the same way to temporarily suppress nematode populations.

III. NATURAL CONTROL

A. Examples of Natural Control

There are many reports of the occurrence of parasites and predators of nematodes in cultivated soils but little quantitative data on their effects on nematode populations. Only in a few cases has the role of naturally occurring antagonists been investigated at more than a superficial level.

1. Decline of Heterodera avenae

The failure of the cereal cyst nematode, *Heterodera avenae* Woll. to increase when susceptible cereals are grown in intensive rotations is the best documented example of naturally occurring control of a plant-parasite nematode. The phenomenon was first reported from England by Gair et al.[11], who followed populations of *H. avenae* in a field where

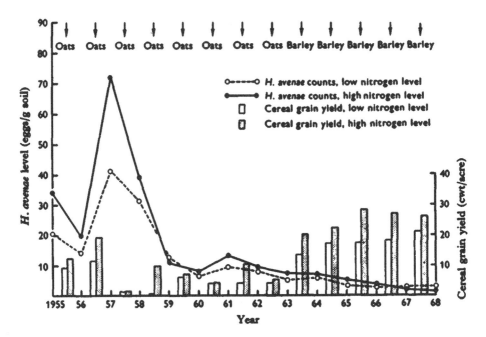

FIGURE 1. Post-cropping levels of *Heterodera avenae* and grain yields when cereals were grown continuously in a field in England. (From Gair, R., Mathias, P. L., and Harvey, P. N., *Ann. Appl. Biol.*, 63, 506, 1969. With permission.)

cereals were grown for 13 years at high and low nitrogen levels. Under both nitrogen regimes, nematode populations peaked 2 years after the start of the experiment and then fell rapidly, the decline continuing for the rest of the experiment (Figure 1). The decline was not the result of a reduction in the capacity of injured root systems to support the nematode because the decline was not arrested when barley, an efficient host of the nematode and a more vigorous crop, replaced oats after 8 years. The cereal cyst nematode also failed to increase under intensive cereal cultivation in other parts of Europe.[12,13] Formalin applied as a drench at 3000 ℓ/ha increased the multiplication of *H. avenae*[15,16] (Figure 2), suggesting that the chemical removed a parasite or competitor which was limiting reproduction.

The phenomenon of decline in cereal cyst nematode populations has been the subject of detailed study by B.R. Kerry and colleagues at Rothamsted in England. Four main species of nematode-parasitic fungi were found in cereal fields infested with *H. avenae*,[17] but *Nematophthora gynophila* Kerry and Crump and *Verticillium chlamydosporium* Goddard were the most widespread. *N. gynophila* invaded cyst-nematode females as they emerged on roots and completely destroyed them within a week. Small, empty cysts were produced when virgin females were infected by *V. chlamydosporium*. Egg producing females also were attacked; infected females produced fewer eggs than normal and many of the eggs were parasitized.

During surveys of cereal fields in which *H. avenae* populations increased or declined, Kerry and Crump[18] found that in both situations, similar numbers of new females were produced (Figure 3A). In soils where nematode multiplication occurred, there was a positive linear relationship between the number of females per root system and the number of new cysts and eggs in soil, whereas in soils where populations declined and females were heavily attacked by fungi, there was no such relationship (Figures 3B, C). Because of the difficulty of extracting fragile, diseased females from soil and the disintegration of cysts, rates of infection of nematodes by fungi, and losses of females generally were underestimated. However, 50 to 60% of the mortality of *H. avenae* females in some field plots were accounted for by *N. gynophila* and *V. chlamydosporium*[20] (Figure 4).

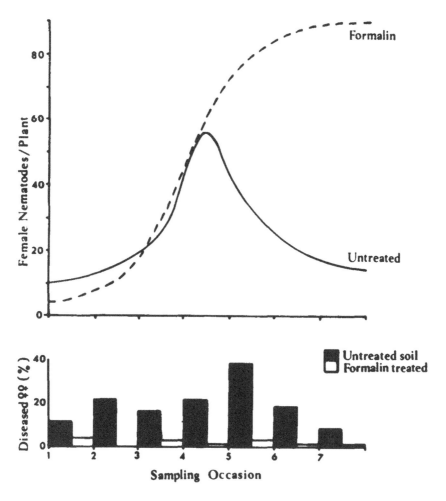

FIGURE 2. Effect of formalin soil drenches (3000 ℓ/ha) on the numbers of females of *Heterodera avenae* on barley roots and parasitism by *Nematophthora gynophila* in pots on 7 weekly sampling occasions. (Figure 5.2 from Brian R. Kerry, "Progress in the Use of Biological Agents for Control of Nematodes," from George C. Papavizas, ed., *Biological Control in Crop Production* (Totowa, N.J.: Allenheld, Osmun, 1981), p. 87. With permission.)

The activity of fungal parasites of *H. avenae* was influenced markedly by soil moisture. More than 90% of females and eggs were killed by these fungi under moist conditions in pots, but levels of parasitism were considerably reduced where soil was kept dry.[16] Fungal parasitism was therefore more severe when conditions were moist during May, June, and July, the period when adult females of *H. avenae* were exposed on the roots. Fungi were less active during dry summers or in free draining soils and this may have accounted for the greater incidence of the nematode in light sandy soils.

2. *Parasitism of Meloidogyne spp. on Peach by Dactylella oviparasitica*

Lovell peach was once widely used as a peach rootstock in California, but because of its susceptibility to root-knot nematodes most growers changed to the *Meloidogyne*-resistant Nemaguard rootstock. In the early 1970s Ferris et al.[21] studied nematode populations in a few of the remaining old orchards on Lovell rootstock and had difficulty finding sites with high populations of root-knot nematodes. Since soil and climatic conditions were suitable for the nematode they speculated that these orchards may have been situated in areas bio-

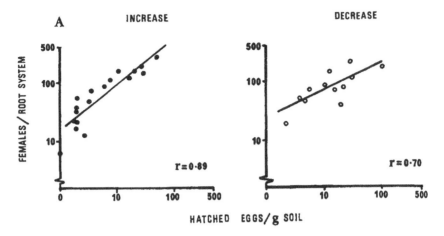

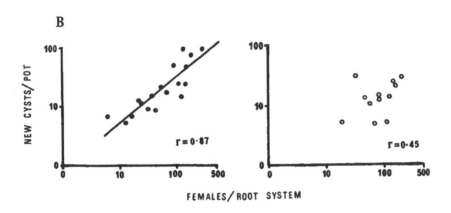

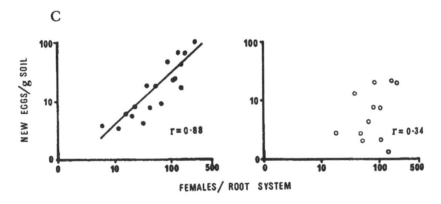

FIGURE 3.   Relationships between (A) number of hatched eggs and number of new females per root system; (B) females per root system and new cysts per pot, and (C) females per root system and new eggs per gram, in soils where *Heterodera avenae* populations increased or decreased under spring barley in pots. (From Kerry, B. R. and Crump, D. H., *Nematologica*, 23, 198, 1977. With permission.)

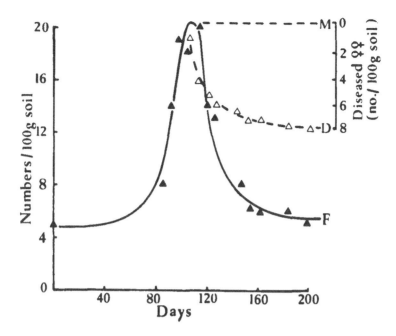

FIGURE 4. Changes in numbers of females and full cysts of *Heterodera avenae* and parasitism by *Nematophthora gynophila* through a growing season on barley. ▲ numbers of females and cysts with contents; △ accumulated numbers of females destroyed by *N. gynophila*. The number of females killed is the difference between the maximum number of females and full cysts (M) and the number of full cysts post harvest (F). Females parasitized by fungi (D) account for approximately 50% of the lost females. (From Kerry, B., *J. Nematol.*, 12, 256, 1980. With permission.)

logically unsuited to *Meloidogyne*. The possibility that the nematode was under natural biological control prompted a search of the area for antagonists and led to the discovery of *Dactylella oviparasitica* Stirling and Mankau, a fungus parasitic in the eggs of the root-knot nematode.[22] *D. oviparasitica* penetrated eggs from appressoria which formed on the surface of the egg, and then proliferated through the egg and eventually destroyed it. Because root-knot nematode eggs were clumped together in masses on the root surface, often all the eggs produced by a female were destroyed.[23,24]

Samples taken throughout the year showed that between 20 and 60% of the eggs were parasitized by *D. oviparasitica*[25] but the actual level of parasitism may have been much higher because the fungus destroyed eggs in less than 9 days at 27°C[26] and some parasitized eggs probably disappeared before being counted. Addition of *D. oviparasitica* to sterilized soil in pots reduced root galling on peach and the number of nematodes in the soil.[25]

Although these observations suggested that *D. oviparasitica* was responsible for the low populations of root-knot nematode on peach trees in California, they did not explain why the fungus had no apparent effect in adjacent vineyards where the nematode and parasite also were present. Greenhouse experiments then showed that *M. incognita* (Kofoid and White) Chitwood produced relatively small egg masses containing a maximum of 350 eggs on peach, whereas egg masses on grape contained more than 1600 eggs. *D. oviparasitica* invariably parasitized most of the eggs in the small egg masses on peach, but on hosts such as grape only about 50% of the eggs were parasitized and many viable eggs remained.[25] Since *M. incognita* produces 1000 to 2000 eggs per egg mass on most hosts, the control achieved by *D. oviparasitica* in peach orchards probably was unique because it depended as much on the inability of the nematode to produce large numbers of eggs on the host plant as it did on the activity of the parasite.

3. Parasitism of Criconemella xenoplax by Hirsutella rhossiliensis

Criconemella xenoplax (Raski) Luc and Raski is one of several factors contributing to peach tree short life in the south-eastern U.S. Observations that nematode populations in some orchards had declined unexpectedly, even when customary weather patterns and farm practices prevailed, led to suggestions that the reduction in nematode numbers may have been caused by parasites and predators. Surveys of South Carolina peach soils then showed that dead *C. xenoplax* with brown heads and distorted, hyphae-filled bodies were often present and occasionally a large proportion of the nematode population was parasitized. *Hirsutella rhossiliensis* Minter and Brady was consistently isolated from these dead nematodes and was able to penetrate and kill juvenile and adult *C. xenoplax* under laboratory conditions,[27] and suppress nematode multiplication in the greenhouse.[28] The impact of *H. rhossiliensis* on the population density of *C. xenoplax* in the field has not been determined, but in the presence of the fungus, nematode numbers generally remained large enough to cause significant injury to peach trees. However, Zehr[29] suggested that the fungus might have been partially responsible for the rapid fluctuations of *C. xenoplax* populations often observed in peach orchards and might be involved in the occasional collapse of nematode populations observed in some orchards.

4. Parasitism of Root-Knot Nematodes by Pasteuria penetrans

Pasteuria penetrans (Thorne) Sayre and Starr, an obligate pathogen of a range of plant-parasitic nematodes including *Meloidogyne, Pratylenchus, Xiphinema,* and *Helicotylenchus* is distributed throughout the world[30] but its role in the population dynamics of nematodes on agricultural crops is not understood. Most reports do not provide useful information on the incidence of parasitism because many authors have incorrectly considered that nematodes with spores attached are parasitized, when penetration into and proliferation within the host are the criteria which should have been used to define infection. Despite the paucity of data, there is little doubt that large numbers of nematodes in cultivated soils are attacked by *P. penetrans,* particularly where a stable monoculture or a perennial cropping system has existed for many years. For example, up to one-third of the *Meloidogyne* females in sugar cane roots in Mauritius and South Africa, and grape roots in South Australia, were infected by the pathogen.[31-33] It is not known whether such levels of parasitism have an impact on nematode populations; in fact, the frequent occurrence of *P. penetrans* in situations where nematode populations were high led Spaull[33] to question its importance in population regulation. Populations of *Meloidogyne* in South African sugar cane fields were higher in fields with *P. penetrans* than in those without and Spaull believed that at the time of sampling the pathogen was not limiting the *Meloidogyne* populations but merely removing surplus individuals. In contrast, old vineyards in South Australia where *P. penetrans* was common tended to have lower populations of *Meloidogyne* than young vineyards where the pathogen was rare.[32] *Meloidogyne* populations increased less in soil from vineyards naturally infested with the pathogen than in similar soils without the pathogen,[34] suggesting that levels of *P. penetrans* in these vineyards may have been sufficient to reduce root-knot nematode populations.

5. Natural Parasitism of Females and Eggs of Cyst and Root-Knot Nematodes

The eggs of cyst and root-knot nematodes are relatively easy to recover from soil because they are aggregated within the body of the female or in an egg mass on the root surface. Hence most of the reports of natural parasitism of nematode eggs have concerned these nematodes. In addition to the previously quoted example of fungal parasitism of *Heterodera avenae* in European cereal fields, reports from many other parts of the world suggest that fungi are often associated with cyst nematodes (Table 1). Some of the published data is difficult to interpret because the extent of disease has been expressed as the number of

Table 1
SOME RECENT REPORTS OF NATURAL FUNGAL PARASITISM OF EGGS, CYSTS, AND FEMALES OF CYST NEMATODES

Author and location	Nematode	Sampling unit	Degree of parasitism	Fungi associated with eggs and cysts	
				Principal parasitic species	Other species
Kerry[17] England	*Heterodera avenae*	Unknown number of cysts from 30 fields	An average of 17% of females killed by *Nematophthora gynophila*. *Verticillium chlamydosporium* was common	*Nematophthora gynophila, Verticillium chlamydosporium, Catenaria auxiliaris, Cylindrocarpon destructans*	—
Bursnall and Tribe[35] England	*Heterodera schachtii*	800 cysts from 7 fields	16% of the cysts contained more than 25% diseased eggs	*Verticillium chlamydosporium, Cylindrocarpon destructans,* a group of black yeasts, an unidentified species with dark, sterile mycleium	15 species, including *Fusarium tabacinum* and 2 *Phoma* spp.
Willcox and Tribe[36] England	*Globodera rostochiensis*	More than 5000 cysts from 19 field populations	No diseased cysts	—	—
	Heterodera avenae	587 cysts from 3 fields	Less than 1% of the cysts contained some diseased eggs	—	*Endogone* sp.
	Heterodera schachtii	741 cysts from 1 field	7% of the cysts contained more than 25% diseased eggs	*Verticillium chlamydosporium*	*Penicillium* sp.
Tribe[37] England and Europe	*Heterodera schachtii*	3164 cysts from 73 populations	17% of cysts were partially or substantially diseased	*Verticillium chlamydosporium, Catenaria auxiliaris, Cylindrocarpon destructans*	Three unidentified species
	Heterodera avenae	177 cysts from 4	13% of cysts were substantially	*Verticillium chlamydosporium*	One unidentified

Table 1 (continued)

SOME RECENT REPORTS OF NATURAL FUNGAL PARASITISM OF EGGS, CYSTS, AND FEMALES OF CYST NEMATODES

Author and location	Nematode	Sampling unit	Degree of parasitism	Fungi associated with eggs and cysts	
				Principal parasitic species	Other species
Nigh et al.[39] California	*Heterodera schachtii*	32 sugar beet fields	10—20% of the eggs parasitized, except in 5 fields where levels of parasitism were greater then 40%	*Fusarium oxysporum, Acremonium strictum*	18 fungi in genera such as *Phoma, Chaetomium, Alternaria, Cephalosporium, Fusarium, Cylindrocarpon,* and *Penicillium*
Morgan-Jones and Rodriguez-Kabana[40] Alabama	*Heterodera glycines*	100 cysts from 1 field	20% of the cysts contained parasitized eggs	*Exophiala pisciphila, Fusarium oxysporum, Fusarium solani*	*Neocosmospora vasinfecta, Phoma multirostrata. Verticillium leptobactrum,* and 3 other common soil species
Morgan-Jones et al.[41] Arkansas, Florida,	*Heterodera glycines*	1000 cysts from 4 sites	Fungi were isolated from 64% of mature cysts	*Fusarium oxysporum, Fusarium solani, Stagonospora heteroderae*	*Exophiala pisciphila, Gliocladium roseum, Neocosmospora vasinfecta, Neocosmospora vasinfecta,* and 21

Reference/Location	Nematode	Sample	Observations	Main fungus	Other fungi
Kerry et al.[43] England	*Heterodera avenae*	Females and cysts from 3 sites	Less than 10% females killed by *Nematophthora gynophila*. *Verticillium chlamydosporium* isolated from 41%, 71%, and 75% of eggs from 3 sites	*Verticillium chlamydosporium*	*Michrodochium balleyi, Gliocladium roseum, Fusarium sp., Paecilomyces sp., Cylindrocarpon destructans, Phoma sp.,* and 6 miscellaneous species
Clovis and Nolan[44] Newfoundland, Canada	*Globodera rostochiensis*	1125 cysts from 5 fields	The predominant fungus (fungus J) was present in 41% of cysts	*Fusarium oxysporum, Coniothyrium fuckelii, Humicola grisea,* unidentified fungus (fungus J)	Fifteen fungi, including *Cylindrocarpon sp., Oidiodendron flavum,* and *Trichocladium* spp.
Volvas and Frisullo[45]	*Heterodera mediterranea*	Unknown number of white females	4-16% of eggs in white females and cysts infected with *Cylindrocarpon destructans*	*Cylindrocarpon destructans*	—
Stirling and Kerry[46] South Australia and Victoria, Australia	*Heterodera avenae*	More than 23,000 females from 375 sites	Approximately 1% of the females were diseased	*Verticillium chlamydosporium, Catenaria auxiliaris,* unidentified sterile fungus	*Cylindrocarpon destructans, Myzocytium sp.,* and *Fusarium* spp.
Gintis et al.[47] Alabama	*Heterodera glycines*	3,374 cysts and females from 1 field	70% of the mature cysts, 50% of the cream-colored cysts, 20% of the white females, and 2% of the sausage-shaped females contained fungi	*Fusarium oxysporum, Fusarium solani, Neocosmospora vasinfecta, Phoma terrestris, Scytalidium fulvum*	*Paecilomyces lilacinus, Paecilomyces variotii, Trichosporium beigelii, Chaetomium coch-*

Table 1 (continued)
SOME RECENT REPORTS OF NATURAL FUNGAL PARASITISM OF EGGS, CYSTS, AND FEMALES OF CYST
NEMATODES

Author and location	Nematode	Sampling unit	Degree of parasitism	Fungi associated with eggs and cysts	
				Principal parasitic species	Other species
Morgan-Jones et al.[48] Columbia	*Heterodera glycines*	900 cysts from 2 sites	Fungi were isolated from approximately 40% of cysts	*Fusarium oxysporum, Fusarium solani*	*Geotrichum candidum, Gliocladium catenulatum, Gliocladium roseum, Paecilomyces lilacinus,* and 13 miscellaneous species
Francl and Dropkin[49] Missouri	*Heterodera glycines*	729 cysts from 5 sites	Chlamydospores of *Glomus* spp. in 9% of cysts	*Glomus* spp., mainly *G. fasciculatum* (no attempts made to detect other fungi)	—

diseased cysts rather than the number of parasitized eggs. Also, fungi found in young females generally can be considered parasitic, but their presence in cysts may indicate a saprophytic rather than a parasitic relationship. Nevertheless, it is apparent that there is considerable natural mortality in most cyst nematode populations. *Globodera rostochiensis* Woll. may be an exception to this general rule because populations in England do not appear to have effective natural enemies.[18,36]

Evidence from numerous surveys suggests that fungi such as *Verticillium chlamydosporium*, *Fusarium solani* (Mart.) Sacc., and *F. oxysporum* Schlecht. are regularly associated with cyst nematodes. A restricted and relatively distinct mycoflora which includes *Geotrichum candidum* Link, *Gliocladium catenulatum* Gilman and Abbot, *G. roseum* Bain., *Paecilomyces lilacinus* (Thom) Samson, *P. variotii* Bain., *Phoma terrestris* Mont., and *Exophiala pisciphila* McGinnis and Ajello also is commonly involved in cyst nematode pathology. Some of these organisms are believed to be associated with the decline of cyst nematode populations or to contribute to nematode control in some situations,[38,39] but further work will be needed to confirm this. Most of them probably are general soil saprophytes with the capacity to colonize cysts and eggs. They may account for some nematode mortality but they do not appear to provide effective nematode control.

Parasitism of *Meloidogyne incognita* eggs by *Dactylella oviparasitica* in California peach orchards already has been mentioned, but other cases of natural fungal parasitism of root-knot nematode eggs have been reported. For example, *Verticillium chlamydosporium* was readily isolated from maturing females and eggs of *M. arenaria* (Neal) Chitwood[50] and approximately one-third of the eggs in small samples of *M. arenaria* and *M. incognita* eggs from peanut and soybean fields in Alabama, U.S. were parasitized by fungi.[51,52] It is apparent that the eggs of endoparasitic nematodes normally are subjected to parasitism, predation, and perhaps antibiosis in the field. However, such effects may not always have a significant impact on nematode population dynamics because these nematodes can sustain high juvenile and egg mortalities without a reduction in their populations.[14,53]

6. Effects of Miscellaneous Parasites and Predators

Other than the examples discussed above, little is known of the importance of parasites and predators of nematodes in agroecosystems. Despite the ubiquity of nematode-trapping fungi, Cooke[54] commented that they had rarely been observed attacking nematodes under natural conditions. Linford and Yap[3] found that nematodes freshly washed from soil had mycelium and traps of nematode-trapping fungi attached. They believed that these fungi captured and destroyed plant-parasitic nematodes in soil but did not attempt to measure the level of predation. Stirling et al.[25] studied soil from peach orchards in California, U.S. which contained several species of nematode-trapping fungi. Although ring traps of *Arthrobotrys dactyloides* Dreschler were observed amongst suspensions of nematodes extracted from those soils, the authors were unable to show that any of the fungi trapped *Meloidogyne incognita* juveniles.

Examples where natural levels of parasitism by parasitic rather then predacious fungi were recorded have largely involved the genus *Hirsutella*. In fields of oil radish in West Germany where low rates of multiplication of *Heterodera schachtii* Schmidt were sometimes observed, up to 90% of the second-stage juveniles were parasitized by *Hirsutella heteroderae* Sturhan and Schneider.[55] This level of disease was similar to that observed in populations of *Criconemella xenoplax* infected by *H. rhossiliensis* (see previously). However, it contrasts with the situation in the cereal fields of southern Australia where the parasitic fungi *H. rhossiliensis* and *Nematoctonus haptocladus* Dreschler were rarely found, and when they did occur they parasitized fewer than 15% of the second-stage larvae of *Heterodera avenae*.[46]

The role of predacious microarthropods in the ecology of plant-parasitic nematodes has rarely been studied in the field. Predacious microarthropods feed voraciously on nematodes in vitro, and since 10^4 to 10^5 springtails (Collembola) and mites per m^2 are commonly found in the top 15 cm of soil,[56,57] it has sometimes been assumed that they have a major impact on populations of plant-parasitic nematodes. However, it is now apparent that levels of predation observed on exposed surfaces in the laboratory substantially overestimate the number of nematodes consumed in soil.[58] Micro-arthropods tend to have omnivorous food habits and may not necessarily target plant-parasitic nematodes when alternative sources of food are available. Also, their spatial distribution may not coincide with that of many plant-parasitic nematodes. Springtails and mites tend to be localized in the uppermost layers of the soil profile and in litter,[56,59,60] whereas plant-parasitic nematodes usually are found throughout the soil profile.

Some predacious nematodes are capable of considerable predacious activity in vitro[61-63] and in soil,[64,65] but there is almost no information on the role of predacious nematodes in the ecology of nematode communities in either natural or agricultural ecosystems. In one of the few studies of predacious and plant-parasitic nematodes in natural soils, fluctuations in populations of *Iotonchus monhystera* (Cobb) Jairajpuri and *Helicotylenchus dihystera* (Cobb) Sher were monitored in a cultivated field and a pasture.[66] Although cyclical changes in predator and prey populations were observed, evidence linking reductions in *H. dihystera* populations with increasing predation by *I. monhystera* was not provided. In a similar study in grassland and arable soils in England, populations of mononchid nematodes could not be correlated with those of other soil inhabiting nematodes.[67]

B. Aspects Requiring Further Attention

1. Detection of Natural Control

The few examples of naturally occurring biological control discussed previously have all been recognized during the last 15 years. There are undoubtedly other examples where the activity of naturally occurring antagonists is sufficient to maintain nematode populations below economic thresholds and concerted attempts must be made to identify them. Nematologists working in the field occasionally find situations where nematode populations are unexpectedly low and crop losses are minimal, and tend to rationalize them as being due to differences in factors such as cultivar susceptibility, soil texture, or cropping history. The possibility that the nematode is under biological control is rarely considered, but it is just these situations where examples of natural control are likely to be found. Since nematodes and antagonists need time to reach a natural equilibrium, old orchards and vineyards, or stable annual cropping systems where intensive rotations have been established for many years, are particularly worthy of investigation.

There is also a need to determine the role of antagonists in typical field situations, where parasitism and predation undoubtedly occur but may not always be sufficient to reduce nematode populations to nondamaging levels. If the various natural mortality factors acting on a nematode population were known and their relative importance understood, it might be possible to use the information to develop control strategies which augment rather than interfere with the natural biological balance.

2. Estimation of Natural Control

If the role of antagonists in the ecology of plant-parasitic nematodes is to be understood, more attention must be given to developing methods of quantifying levels of parasitism and predation. The number of nematodes killed by natural enemies in field soil has been estimated most commonly by regularly extracting and counting diseased and healthy nematodes. This method has proved most useful for studies of fungal parasitism in root-knot and cyst-nematode eggs because they are readily recoverable from soil, but is less useful for studies of parasitism on vermiform nematodes and nonaggregated eggs. Moribund nematodes can be extracted from soil using sugar flotation techniques,[25,27,46] but it is not known whether they are extracted with the same efficiency as healthy nematodes.

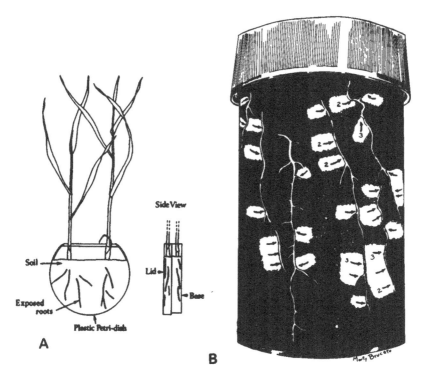

FIGURE 5.   Observation chambers for evaluating parasitism and predation on cyst nematode females. (A) Petri-dish containing a cereal seedling infected with *Heterodera avenae*. (From Crump, D. H. and Kerry, B. R., *Nematologica*, 23, 399, 1977. With permission.) (B) Roots from potato tuber infected with *Globodera rostochiensis*, showing number and distribution of females. (From La Mondia, J. A. and Brodie, B. B., *J. Nematol.*, 16, 113, 1984. With permission.)

Baiting techniques, in which nematodes are added to soil and later recovered and examined for parasitism have proved useful in some studies. For example, parasitism of *M. incognita* eggs by *Dactylella oviparasitica* was studied by burying egg masses in the field in containers which allowed ingress of fungi but facilitated the recovery of eggs.[68] A simple agar disc technique has also been used to sample soil for fungi capable of colonizing nematode eggs and for estimating their level of activity.[69] During studies of parasitism and predation on females and eggs of root-knot and cyst nematodes, plants infected with nematodes were grown in the test soil, and females and eggs examined as they developed.[18,25,46] Observation chambers have been developed to facilitate such observations (Figure 5).

The application of insecticides which inhibit or kill natural enemies often leads to resurgence of an insect population after an initial period of suppression.[4] Such treatments have been used by entomologists to determine the degree of natural suppression in an insect population, but have been used infrequently by nematologists. Formalin applied as a soil drench to cereal fields in England caused populations of *Heterodera avenae* to increase[15] and it was later shown that the chemical inhibited parasitism by *Nematophthora gynophila* and *Verticillium chlamydosporium*. The increase in nematode populations was considered a measure of the suppression by these fungi.[16,72] In a pot experiment with soil from a vineyard in South Australia, the soil was autoclaved or treated with ethylene dibromide or 1,3 dichloropropene and then inoculated with root-knot nematodes. Nematodes multiplied more rapidly in the autoclaved soil than in the soil treated with nematicides, possibly because the naturally occurring antagonist *Pasteuria penetrans* was killed by autoclaving but not by the nematicides.[34] There is undoubtedly potential for further work using such techniques, because a range of insecticides, miticides, fungicides, and antibiotics are available to selectively eliminate or inhibit various components of the soil microflora and fauna.

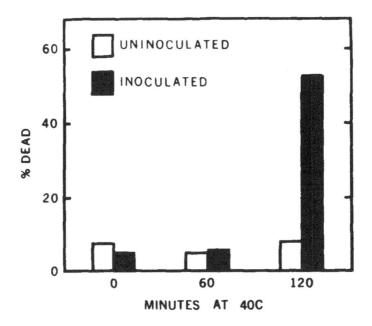

FIGURE 6.   Mortality of *Criconemella xenoplax* exposed to sublethal heat (40°C for 0, 60, or 120 minutes) prior to inoculation with *Hirsutella rhossiliensis*. (From Jaffee, B. A. and Zehr, E. I., *Phytopathology*, 72, 1381, 1982. With permission.)

The main problem involved in estimating the level of natural control is the rapid disintegration of diseased individuals. For example, the contents of *Meloidogyne* eggs were completely consumed by *Dactylella oviparasitica* in 9 days at 27°C[26] while females of *Heterodera avenae* attacked by *Nematophthora gynophila* were unrecognizable after 4 days at 13°C.[73] Estimates of the number of nematodes consumed by predators are even more difficult to obtain because the individuals preyed upon usually disappear without trace.

Although an organism may be isolated from or found associated with a nematode or nematode egg, this does not necessarily indicate a parasitic relationship. Thus the nature of the relationship between suspected parasites and nematodes must be clarified before attempts are made to determine the impact of these organisms on nematode populations. Some of the many fungi isolated from *Heterodera* and *Globodera* cysts probably obtain their nutrients from the mucilage produced when internal tissues of the female break down, rather than from eggs within the cyst. Others are not aggresive parasites of eggs on agar[74] and may exist largely as saprophytes in soil. Sometimes it is difficult to determine whether an organism is parasitizing nematodes or invading dead or unhealthy individuals. Changes brought about by stress or sublethal concentrations of nematicides can predispose nematodes to attack by *Catenaria anguillulae* Sorokin,[75,76] while Jaffee and Zehr[27] suggested that *Criconemella xenoplax* stressed by sublethal temperatures, starvation, dry soil conditions, or sublethal concentrations of nematicides may be more susceptible than healthy nematodes to parasitism by *Hirsutella rhossiliensis* (Figure 6).

III. INTRODUCTION OF ANTAGONISTS

By the time interest in plant-parasitic nematodes began to develop and grow during the 1930s and 1940s, there were several examples of the successful introduction of a natural enemy into a new environment to control an insect pest or a weed.[4] Some of the early nematologists were undoubtedly influenced by these successes and the literature on biological control of nematodes has tended to be dominated by the so called "classical" approach to biological control, in which parasites and predators were introduced into soil in an attempt to control nematodes.

A. Examples

1. Nematode-Trapping Fungi

Until a few years ago, attempts to achieve nematode control through the introduction of antagonists almost without exception involved the nematode-trapping fungi. Interest in these fungi was stimulated by the work of Linford et al.[77] who found that root-knot nematode populations in soil from pineapple fields declined during the decomposition of organic materials, and suggested that the decline could be attributed to the activity of nematode-trapping fungi. At the same time, Dreschler's prolific descriptive studies on the nematode-trapping fungi made biologists aware of their diversity and ubiquitous distribution, and aroused interest in their unique trapping mechanisms.

During the last 50 years there have been numerous attempts in many countries to control nematodes by introducing nematode-trapping fungi, but the results have rarely been encouraging (Table 2). Although pot experiments and small-scale field trials sometimes have yielded promising results, the high level of nematode control required in modern farming systems has never been consistently achieved on a field scale. Some commercial preparations of nematode-trapping fungi have been marketed, but the products have never been used widely because of quality control problems and inconsistent performance. *Arthrobotrys robusta* Duddington, commercially formulated as Royal 300®, reduced populations of *Ditylenchus myceliophagus* Goodey and increased yields of the cultivated mushroom, *Agaricus bisporus* (Lange) Sing.,[88] while Royal 350®, a similar product containing *Arthrobotrys superba* Corda, gave adequate control of root-knot nematode on tomato provided it was not used in situations where nematode populations were high.[85,89]

With the benefit of hindsight, it is relatively easy to speculate on the reasons for the inconsistent performance of the nematode-trapping fungi. In experiments where mycelium was added to soil, establishment was likely to have been poor because of competition from the soil microflora. The introduced inoculum would have provided a substrate for a range of soil microorganisms and since it was often added at rates equivalent to several tonnes of fungal material per hectare, any nematode control observed could have been due to the products of microbial decomposition rather than trapping activity. The fungistatic nature of soil is now well recognized[90] and nematode-trapping fungi introduced as spores may not have been able to germinate and become established because of severe fungistatic and lytic effects.[91,92]

Most experiments with the nematode-trapping fungi have involved the use of large quantities of organic amendments and because it was assumed that these materials stimulated fungal activity, predation by fungi usually was considered responsible for any nematode control observed. It is now apparent that a complex series of changes take place in soil during the decomposition process and that many chemicals are produced which have deleterious effects on nematodes. Some are nematicidal, others change nematode behavior by affecting egg hatch, motility, or the egg laying capacity of females, while others reduce the susceptibility of roots to invasion by nematodes.[93-101] Careful analysis of experiments where nematode-trapping fungi have been introduced into soil with a variety of organic substrates and amendments shows that there is little direct evidence that the fungi were responsible for the nematode control observed. Few attempts were made to reisolate the fungus after its introduction, or to measure changes in the number of fungal propagules and predacious activity during the course of experiments. It is most likely that nematicidal products produced during the decomposition cycle caused the reductions in nematode populations while improvement in plant growth may have been due to the beneficial effects of organic matter on plant nutrition and soil structure. It is perhaps unfortunate that nematode-trapping fungi have

Table 2

SOME EXPERIMENTS WITH NEMATODE-TRAPPING FUNGI FOR THE CONTROL OF PLANT-PARASITIC NEMATODES

Fungus, substrate and application rate	Amendment[a]	Nematode and host	Soil and plot size	Summary of results	Ref.
Arthrobotrys oligospora, *A. musiformis*, *A. candida*, *A. thaumasia*, and *M. ellipsosporium* grown on sugar cane bagasse. 0.13% w/v	Nil	*Meloidogyne* sp. on pineapple	Sterilized soil in pots	*M. ellipsosporium* increased leaf length, root length and root weight but did not affect the number of galls per root system. Other fungi had no effect on galling or plant growth.	3
Mycelium of *Arthrobotrys thaumasia* grown in broth culture. 7 t/ha	± 20 t/ha bran	*Heterodera schachtii* on sugar beet	Field soil in microplots	The fungus alone increased yield but had no effect on numbers of cysts and eggs. Bran alone increased yield and also reduced cyst numbers but did not affect egg numbers. Fungus + bran produced similar results to the fungus or bran alone.	78
Macerated mycelium of *Arthrobotrys conoides* and *A. dactyloides* grown in broth culture[b]	Nil	*Meloidogyne incognita* on tomato and okra	Quartz sand in pots	Similar level of galling on fungus-inoculated and control plants.	79
Arthrobotrys conoides, *A. dactyloides*, *A. arthrobo-* ...	± 2.3% v/v sugar beet pulp	*Meloidogyne incognita* on tomato	Steam-sterilized soil in pots	*M. ellipsosporium* and *A. conoides* added to amended soil and *A. thauma-* ... in unamended soil increased sur- ...

lipsosporium grown in broth culture or as powdered dry spores[b]					
Arthrobotrys robusta and *A. candida* grown on nutrient medium in vermiculite and then air-dried. 0.1% w/w	± 0.5% w/w bran	*Heterodera avenae* on oats	Field soil in pots	Bran or fungi + bran gave a similar reduction in the number of nematodes in roots. In the absence of bran, the fungi had no effect on nematode numbers in roots.	80
Spores or dried mycelium of *Arthrobotrys robusta* and *A. candida*. 33 × 10⁶ spores/m²	5 t dry matter/ha bran, farmyard manure or chopped cabbage leaves	*Globodera rostochiensis* on potato and *Heterodera gottingiana* on pea	Field soil in small pots	Bran or farmyard manure combined with the fungi failed to increase yield. Chopped cabbage sometimes increased yields and the superimposition of the fungus did not result in additional improvement.	
Arthrobotrys thaumasia mycelium grown in broth culture. 3 t/ha (experiment 1), or 0.5 t/ha (experiment 2)	± 12 t/ha grass mowings (experiment 1), ± 4 t/ha chopped cabbage (experiment 2)	*Heterodera avenae* on oats	Field soil in plots approx 0.5 m²	Green manure reduced the number of nematodes in roots in experiment 1. The fungus alone reduced nematode numbers in both experiments. The addition of green manure increased the degree of control produced by the fungus in experiment 2 but not in experiment 1.	81
Arthrobotrys musiformis grown on oat hulls. 62.5 and 125 t/ha	Control plots received 125 t/ha oat hulls	*Radopholus similis* on grapefruit	Soil in pots	Plants treated with the oat hull-fungus mixture had more nematodes in roots than controls receiving only oat hulls.	82
A. musiformis grown in fermentation tanks. The	Nil	*Radopholus similis* on orange	Field soil in plots around mature or-	No evidence of nematode control by the fungus during the first 8 months	

Table 2 (continued)
SOME EXPERIMENTS WITH NEMATODE-TRAPPING FUNGI FOR THE CONTROL OF PLANT-PARASITIC NEMATODES

Fungus, substrate and application rate	Amendment[a]	Nematode and host	Soil and plot size	Summary of results	Ref.
Spore preparations of a mixture of *Arthrobotrys kirkhizica, A. dolioformis,* and *A. pravicovii* at concentrations ranging from approx. 0.01%–1.1% w/w	20% v/v rotted manure	*Meloidogyne* sp. on an unidentified crop	Field soil in pots	Spore preparations decreased the number of galls per plant by 28-75%.	83
Spore preparations of the above fungi at 1.5 t/ha	± 30 t/ha manure	*Meloidogyne* sp. on tomato	Field soil in 4 m² plots	The preparation of predacious fungi did not affect the number of infected plants but yield increased by 7% (non-replicated experiment).	
Arthrobotrys conoides grown on vermiculite soaked with nutrients. 3100–12400 l/ha (microplots), 7% v/v (pots)	±5.57 t/ha green alfalfa (two microplot experiments); ± 20% w/v chopped alfalfa (two pot experiments)	*Meloidogyne incognita* on maize	Methyl bromide-treated soil in microplots; steam sterilized sand-soil mixture in pots	*A. conoides* and/or alfalfa mulch, both alone or in combination, reduced nematode numbers and root galling in all experiments. These treatments also increased plant growth or yield in all but one microplot experiment.	84
Arthrobotrys superba grown on autoclaved	Nil	*Meloidogyne* sp. on tomato	Steamed soil mix in pots; field soil in 25 m² plots	All concentrations of *A. superba* reduced numbers of *Meloidogyne* females in roots of potted plants. The	85

Organism/formulation	Rate	Nematode/host	Conditions	Results	Ref.
Liquid or granular formulations of *Arthrobotrys amerospora*	Nil	*Meloidogyne incognita* on soybean (pot experiment)	Sterilized fine sand in pots	A. amerospora had no effect on nematode populations or plant growth	86
		Belonolaimus longicaudatus, Haplolaimus galeatus, and *Paratrichodorus christiei* on maize (one field experiment) and the above nematodes plus *M. incognita* on okra (2 field experiments)	Field soil in 17 m² plots	On maize, A. amerospora did not affect nematode populations, plant growth, or grain yield. In experiments with okra, A. amerospora had no effect on nematode numbers or root-galling caused by *M. incognita*.	
Monacrosporium ellipsisporium grown on wheat kernels, 1, 5, or 10 g colonized substrate per pot (0.07, 0.35, 0.7% w/w)	Appropriate amounts of noncolonized wheat kernels added to all pots (treatments and controls) to bring total amount wheat kernels to 10 g per pot (0.7% w/w)	*Meloidogyne incognita* on tomato	Field soil in pots	Galling was reduced by treatments containing 5 and 10 g fungus per pot. The reduction in plant damage was associated with a reduction in larvae in soil and females in roots.	87
5 or 50 g colonized substrate per transplant hole (field experiment)	Appropriate amounts of noncolonized wheat kernels added to treatments and controls to bring total amount wheat kernels to 50 g per transplant hole.		Field experiment (plot size or volume of transplant hole not given)	Plants in all treatments were heavily galled. Plant growth parameters and galling in fungus treated plots did not differ significantly from controls.	

a 1% w/w is approximately equivalent to 20 t/ha incorporated to a depth of 15 cm.

b Application rate not given.

achieved a status as biological control agents that they may not deserve. A major part of the work on biological control of nematodes has been directed towards these organisms and their subsequent failure has produced a generation of nematologists skeptical of the potential for biological control.

Although results with the nematode-trapping fungi have been disappointing, it may be premature to dismiss them as biological control agents because so few species have been studied in detail. The nematode-trapping fungi exhibit tremendous diversity in their modes of nutrition and some species will almost certainly be more effective antagonists then others. Cooke[102,103] studied a range of nematode-trapping fungi and found that ring-forming species and those forming adhesive branches or knobs grew poorly on corn-meal agar and had poor competitive saprophytic ability, but tended to produce traps spontaneously and be efficient predators. In contrast, species with reticulate traps grew more rapidly in culture and had good competitive saprophytic ability, but did not produce traps spontaneously or decrease nematode populations. Although the most efficient predators tend to lose the capacity to compete as saprophytes in soil,[104] it is essential that predacious rather than saprophytic species are chosen if biological control with nematode-trapping fungi is to be successful. It is unlikely that predacious fungi can maintain an effective screen of traps in soil for long enough to control sedentary endoparasitic nematodes such as *Meloidogyne*, but species such as *Monacrosporium lysipagum* (Dreschler) Subram. and *M. ellipsosporium* (Grove) Subram., which can invade egg masses and prey on juveniles as they hatch[87,105] are worthy of further investigation. Also, the nematode-trapping fungi are more likely to be effective against ectoparasitic nematodes and species such as *Tylenchulus semipenetrans* Cobb, whose juveniles migrate in the rhizosphere and feed ectoparasitically before becoming established at a permanent feeding site. It is surprising that the predacious fungi rarely have been tested against such nematodes.

2. Endoparasitic Fungi

The endoparasitic fungi are largely dependent on nematodes for their nutrition and produce only limited mycelial growth outside the host. Species such as *Nematophthora gynophila* and *Catenaria auxiliaris* (Kuhn) Tribe, which are aggressive parasites of some cyst nematodes in moist environments, are obligate parasites, while many other species grow slowly on standard mycological media. Because of the difficulties involved in culturing the endoparasitic fungi, there have been few attempts to utilize them for nematode control. Nematode numbers in sterile sand were reduced when conidia of *Nematoctonus concurrens* Dreschler and *N. haptocladus* were added.[106] However, nematodes were not affected by comparable treatments in nonsterile soil because conidia suffered severe mycostatic and lytic effects (Figure 7), and when germination was successful the germlings lysed. The susceptibility of *Nematoctonus* to fungistasis led the authors to question whether even the massive addition of spores to soil would result in a reduction in nematode populations.

In two greenhouse experiments, *Hirsutella rhossiliensis* suppressed populations of *Criconemella xenoplax* when introduced into autoclaved soil,[28] but more recent work suggests that fungistatic effects may limit its usefulness as a biological control agent. Spores germinated in nonsterile soil but their germ tubes were short and malformed compared with those produced in sterile soil (Table 3). The fungus also lacked good competitive saprophytic ability; it was able to colonize substrates such as dead nematodes or wheat seeds in sterile soil but could not compete with other soil fungi for these substrates in nonsterile soil.[107] Despite these problems, *H. rhossiliensis* has been successfully introduced into the field by adding the fungus to nematode-infested soil containing peach seedlings before the seedlings were transplanted to the field.[29]

Recently, the effectiveness of *Meria coniospora* Dreschler in controlling root-knot nematodes on tomato was compared in sterile and nonsterile soil in pots.[108] Although the fungus

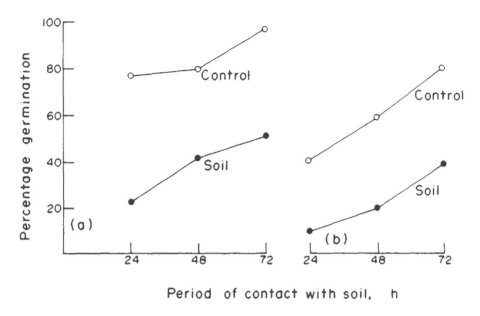

FIGURE 7. Germination of conidia of (a) *Nematoctonus concurrens* and (b) *N. haptocladus* in the presence or absence of nonsterile soil. (Reprinted with permission from *Soil Biol. Biochem.*, 6, Guima, A. Y. and Cooke, R. C., 1974, Pergamon Journal, Ltd.)

Table 3
GERMINATION AND GERM TUBE LENGTH OF
***HIRSUTELLA RHOSSILIENSIS* SPORES INCUBATED**
ABOVE STERILE AND NONSTERILE SOIL FOR 20
AND 40 HR

Soil	20 Hr incubation		40 Hr incubation	
	Germination (%)	Germ tube length (μm)	Germination (%)	Germ tube length (μm)
Sterile	89 ± 8	18 ± 2	93 ± 4	128 ± 13
Nonsterile	91 ± 6	16 ± 1	93 ± 8	33 ± 5

Note: Values are means ± SD of four replications.

From Jaffee, B. A. and Zehr, E. I., *J. Nematol.*, 17, 344, 1985. With permission.

reduced root galling, the value of the data is questionable because most of the nematodes used in the experiment were not infective.

3. Fungi Parasitic in Eggs

Interest in fungi parasitizing females and eggs of endoparasitic nematodes has increased rapidly in recent years. Much of this interest has been directed towards *Paecilomyces lilacinus*, a fungus which aggressively parasitizes nematode eggs on agar[109,110] and provides some nematode control in greenhouse tests.[51] Several groups claim to have controlled root-knot and cyst nematodes by introducing *P. lilacinus* into the field,[111-116] but in many instances the published data is insufficient to fully evaluate the experiments (Table 4). As with much of the work on the biological control of nematodes, there is a lack of convincing evidence that the introduced fungus was responsible for the observed reductions in nematode populations. In most cases, the effects of *P. lilacinus* could not be separated from those of the

Table 4

SOME FIELD EXPERIMENTS WITH *PAECILOMYCES LILACINUS* FOR THE CONTROL OF PLANT-PARASITIC NEMATODES

Substrate for fungus and application rate	Treatments	Nematode/Host	Results	Ref.
Potato dextrose agar (PDA) 2 Petri plates/m²	Fungus + PDA Control (no treatment)	*Meloidogyne incognita* on potato	Fungus + PDA caused a small but significant reduction in root galling index. No data on nematode populations were given but at the end of the experiment 54.7% of the eggs were destroyed.	111, 116
PDA. Application rate not given	Fungus + PDA Control (no treatment)	*Meloidogyne incognita* on successive crops of potato, bean, and potato	No data presented. Fungus + PDA reduced root and tuber galling index in the first two crops but not in the third crop.	112
Substrate and application rate not given	Fungus + substrate Control (no treatment)	*Meloidogyne incognita* on tomato	Fungus + substrate reduced galling but did not affect yield.	114
Chopped water lily leaves 1.75 t/ha	*Experiment 1* Fungus + water lily leaves Tubers dipped in fungal spores Control (no treatment) *Experiment 2* Fungus + water lily leaves Tubers dipped in fungal spores Combination of above Control (no treatment)	*Globodera rostochiensis* on potato	Treatments containing the fungus reduced cyst counts by 41-65%. Two of the eight treatments containing the fungus significantly increased yield.	113

Autoclaved wheat seeds 45.6 g/cm² reported, but probably 45.6 g/m² (0.456 t/ha)	Fungus + wheat seeds Control (fungus + wheat seeds, autoclaved) Control (wheat seeds) Control (no treatment)	*Meloidogyne javanica* on tobacco	Plants in control plots and in plots treated with the fungus yielded poorly, and had badly galled roots.	117
Autoclaved oats 0.5 and 1% w/w	Fungus + oats (3 isolates of fungus) Control (uncolonized oats) Control (no treatment)	*Meloidogyne arenaria* on squash	Uncolonized oats reduced galling compared with the untreated control. Only one fungal isolate further reduced galling at both application rates.	118

Table 5
PLANT GROWTH AND GALLING CAUSED BY
MELOIDOGYNE INCOGNITA IN THE PRESENCE
OR ABSENCE OF *PASTEURIA PENETRANS* SPORES

Soil	Dry weight		Gall rating (0—25)
	Tops (g)	Roots (g)	
Infested with *P. penetrans*	5.3 a	2.6 a	6.5 a
Infested with *P. penetrans*, steam sterilized	3.1 b	2.5 a	20.5 b
Not infested with *P. penetrans*	3.4 b	1.7 a	22.7 b

Note: Means followed by the same letter do not differ significantly, P = 0.05.

Modified from Mankau, R., *J. Invertebr. Pathol.*, 26, 338, 1975. With permission.

substrate on which the fungus was grown, the fungus was not reisolated from treated soil, and fungal populations compared with those in untreated soil, and no attempt was made to assess the level of parasitic activity during the course of the experiments. Such details are important because oats used as a food base for *P. lilacinus* improved plant growth and provided significant control of *Meloidogyne arenaria* while *P. lilacinus* gave little or no additional control.[51,118]

4. Bacteria

Pasteuria penetrans is an obligate parasite of many important plant-parasitic nematodes,[30] but most interest has been centered on populations which attack root-knot nematodes. Recent studies of the life cycle and mode of action of *P. penetrans* have shown that the pathogen is detrimental to second-stage juveniles in soil and also to parasitic stages of the nematode in roots. When spores attach to second-stage juveniles their motility is affected[119] and they often fail to infect roots.[34,120] The number of spore-encumbered nematodes unable to invade roots increases as the spore concentration in soil increases, and plants grown in soil containing high numbers of spores may be virtually free of galls.[34] In such situations, those spore-encumbered juveniles which are able to establish permanent feeding sites in the root usually become infected by the pathogen. Since at least one spore must germinate to initiate infection and not all spores are viable, as many as five spores may be needed to ensure infection.[34] Once infection occurs, the pathogen proliferates throughout the body of the developing nematode and prevents reproduction.[121-125]

Although *P. penetrans* has many of the attributes required by a successful biological control agent, it has not been widely applied to soil in an attempt to control nematodes because it has proved difficult to culture in large quantities. *P. penetrans* gave promising results against *Meloidogyne incognita* and *Pratylenchus scribneri* Steiner in pot experiments (Table 5), but the spore-infested soil used as a source of the pathogen took several years to produce and was inconvenient to transport and handle in the field. The development of a mass production technique in which roots containing large numbers of *Meloidogyne* females infected with *P. penetrans* were air-dried and finely ground to produce a light, easily handled powder,[126] enabled more extensive testing of the pathogen and confirmed its potential as a biological control agent against root-knot nematodes. When dried root preparations laden with spores were incorporated into soil at 212 to 600 mg/kg soil, the number of *M. javanica* (Treub) Chitwood juveniles in soil and the degree of galling caused by the nematode was

FIGURE 8. Roots of tomato grown in field soil infested with *Meloidogyne javanica* and treated with 600 mg of a dried-root preparation of *Pasteuria penetrans* per kg dry soil (left) and grown in untreated soil (right). (From Stirling, G. R., *Phytopathology*, 74, 58, 1984. With permission.)

reduced significantly (Figure 8). *P. penetrans* incorporated into microplots in which tobacco, soybean, and winter vetch were planted in sequence in successive years reduced yield losses caused by *M. incognita* by 23 to 55% for all crops except the first soybean crop.[127]

Since *P. penetrans* can prevent second-stage juveniles of *Meloidogyne* from invading roots, it is interesting to speculate on the concentration of spores likely to be needed in soil at planting to provide control of root-knot nematodes in annual crops. In the laboratory, *Meloidogyne* juveniles added to soil containing 0, 10^3, 10^4, 10^5, and 10^6 spores per gram soil had an average of 0, 0.2, 0.9, 7.7, and 28.6 spores attached per nematode when extracted 24 hr later.[128] Since juveniles probably spend at least 24 hr in soil before locating a root, and the infectivity of nematodes with more than 20 spores attached is reduced,[34,120] spore concentrations of 10^5 to 10^6 spores/g soil should be sufficient to prevent some juveniles from infecting roots. Since those juveniles able to invade roots would probably carry enough spores to ensure that they did not reproduce, a degree of nematode control could be expected at these spore concentrations. The results of field experiments with dried root preparations of *P. penetrans*[34] tend to support these predictions because plants growing in soil containing 424 to 600 mg dried root preparation per kilogram of soil were almost free of galls. Since the spore concentration in the inoculum, as estimated from the number of infected *Meloidogyne* females likely to have been present in each dried root system, was believed to have been 10^8 to 10^9 spores/g,[126] this suggests that control was obtained with spore concentrations of 0.5 to 5 × 10^5 spores/g soil.

Although predictions about the spore concentration needed to control root-knot nematodes in annual crops can only be made by extrapolating from limited data, it appears that *P. penetrans* acts as a biological nematicide when incorporated into soil at concentrations of 10^5 to 10^6 spores/g soil. It is not unreasonable to expect that such large quantities of inoculum could be produced by in vitro culture techniques. *P. penetrans* spores are found in the bodies of infected *Meloidogyne* females at concentrations of approximately 10^{10} spores/mℓ nematodes, and if the pathogen could be grown with the same efficiency in liquid culture the required spore concentrations in the top 15 cm of soil would be achieved with application rates of 20 to 200 ℓ/ha. It is possible that inundative releases of *P. penetrans* may also be useful for temporarily suppressing root-knot nematodes in perennial crops. However, further work on the population dynamics of nematode and pathogen will be needed to determine whether the pathogen has the capacity to sustain itself in such cropping systems and is able to permanently maintain nematode populations at levels below economic thresholds.

Apart from the difficulties involved in developing suitable mass production techniques, other problems will have to be solved before *P. penetrans* is successfully commercialized. Recent evidence suggests that *P. penetrans* exists in nature as host-specific populations, each with a limited host range. Spores collected from root-knot nematodes in one location, for example, do not always attach to and infect *Meloidogyne* populations from other locations.[129] Thus it may be necessary to select populations of the pathogen with a wide host range, or to mix several populations together to counter the diversity in root-knot nematodes usually found in the field.

5. Predacious Nematodes and Microarthropods

Because of the problems involved in mass producing predacious microarthropods and nematodes there have been few attempts to introduce them into field soil for nematode control. Most of the experiments with the predacious microarthropods have been carried out in the laboratory and greenhouse, and have been designed to determine whether particular species are able to prey on nematodes and to quantify the number of nematodes consumed. Numerous species of springtails and mites feed on nematodes or their eggs in culture chambers or Petri dishes,[46,56,68,130-134] while predation on eggs and females of root-knot and cyst nematodes has been observed in pots and observation chambers.[46,135] The results of experiments in small containers suggested that the astigmatid mite *Caloglyphus* sp. could feed on nematodes in the field as well as under laboratory conditions.[133] Sharma[132] compared numbers of *Tylenchorhynchus dubius* (Bütschli) Filipjev on rye seedlings in small glass jars containing no predators with nematode numbers in similar jars inoculated with several species of mites and springtails. In most of the treatments inoculated with microarthropods, nematode numbers were reduced by at least 50%. However, the differences in nematode populations may not have been due totally to predation, because root feeding by nematodes and springtails may have caused differences in root growth.

Some predacious nematodes can consume considerable numbers of nematodes in laboratory cultures[62-64,136] but evidence from pot experiments suggests that they do not have marked effects on nematode numbers in soil. Populations of *Tylenchulus semipenetrans* were not reduced by *Thornia* sp.,[137] while *Prionchulus punctatus* Cobb and *Mylonchulus sigmaturus* Cobb did not have a major effect on populations of ectoparasitic nematodes and *T. semipenetrans*, respectively.[64,65] In contrast to these results, *P. punctatus* alone or in combination with *Mononchus aquaticus* Coetzee and *Labronema* sp. significantly reduced galling caused by *Meloidogyne incognita* on tomato.[65]

Many of the predacious nematodes and predacious microarthropods are unlikely to successfuly regulate a population of plant-parasitic nematodes because they are omnivorous and their population densities are not related to nematode density. However, there is increasing evidence that some predators exhibit preferences for certain types of prey, the suitability of particular nematode species being determined by factors such as their size, motility, and the thickness and degree of annulation of their cuticles.[63,64,134,138] Further studies of some of these less polyphagous species may be justified but the difficulties involved in mass culturing predacious animals should not be underestimated. For example, it may be possible to culture predators in the family Diplogasteridae by modifying methods developed for some of the rhabditoid nematodes used as biological control agents against insect pests.[139] However, even if weekly production levels of 2×10^{10} nematodes per person were attained, it would take 2 weeks for an individual to produce enough nematodes to inoculate an area of 10 ha to a depth of 15 cm with approximately 3000 nematodes per kg soil.

B. Aspects Requiring Further Attention

1. The Search for Better Antagonists

Although an impressive array of parasites and predators of nematodes are discussed in previous chapters of this book, probably no more than a small proportion of the organisms capable of antagonism towards plant-parasitic nematodes have been described. For example, the bacterial pathogens, viruses, and endoparasitic fungi are almost certainly under represented. Most of the known antagonists have been found by chance and identified in soil using a limited range of techniques. For example, many of the nematophagous fungi were discovered by randomly collecting small quantities of soil or organic matter and plating it onto weak nutrient media in Petri dishes. Although such methods have proved invaluable, they tend to favor aggressive predacious fungi and endoparasites of free-living nematodes. Thus, within the fungi, fast-growing, heavily sporulating, saprophytic species probably are over represented compared with slow-growing, host-specific endoparasitic species. The recent discovery of several new fungal parasites of females and eggs of root-knot and cyst nematodes[22,47,73,140,141] shows that antagonists of particular plant-parasitic nematodes can be found by searching in a directed manner. Useful antagonists may be found in any random soil sample, but they should be specifically sought in areas where nematode damage is minimal, where numbers of a particular plant-parasitic nematode have declined, or where high nematode populations do not develop despite the presence of a suitable soil environment and a susceptible host. It may be worthwhile searching the center of origin of some of the world's most important plant-parasitic nematodes, because similar tactics have proved useful in identifying natural enemies of some insect pests and weeds.

In the past, most of the antagonists considered to have potential as biological control agents against nematodes have been parasitic or predacious. However, organisms with other mechanisms of action should also be considered in the search for better antagonists. Lysed, shrivelled, coagulated, or decaying eggs free of fungal hyphae are often found in *Heterodera* cysts,[142] and it has been suggested that these deleterious effects might be due to the production by soil microorganisms of diffusible toxic metabolites which cause physiologic or metabolic disturbances in nematodes or their eggs.[110,142-144] Many fungi and bacteria produce nematoxic compounds in vitro[145-159] and some of these compounds are highly nematicidal. For example, avermectins produced by a species of *Streptomyces* are effective at concentrations ten times less than are needed for some commonly used nematicides.[160] Although these compounds are not necessarily produced in soil at concentrations sufficient to be detrimental to nematodes, this possibility should be investigated because antibiosis has been neglected as a possible mechanism of biological control of nematodes. Another group which should be given further consideration are the vesicular-arbuscular mycorrhyzae. Although they can be weakly parasitic on nematodes,[49] they more commonly promote nematode tolerance in plants[161] and might be used to offset the damage caused by nematodes.

In the past, little consideration has been given to the life cycle of particular groups of plant-parasitic nematodes and the types of antagonists most likely to succeed against them. Many of the early attempts to achieve biological control of root-knot and cyst nematodes were doomed to failure because they involved the use of antagonists which attacked free-living, second-stage juveniles in soil. This stage of the nematode is motile, dispersed, and transient, and probably is the most difficult stage in the life cycle for an antagonist to locate and kill in large numbers. In fact, on some *Meloidogyne*-susceptible host plants an organism preying on second-stage juveniles in soil may never make contact with its prey because succeeding generations of juveniles reinfect galled tissue without ever migrating through soil. Since *Meloidogyne* females are also relatively immune to parasitism and predation because they are protected by the root, the egg is obviously the stage in the life cycle most vulnerable to antagonism. Eggs are located on the root surface and even under ideal environmental conditions they take at least 10 days to develop and hatch. Because they are aggregated in an egg mass, an effective antagonist established in that vicinity could be expected to eliminate all the eggs produced by an individual nematode. This is an important consideration because *Meloidogyne* females produce as many as 2000 eggs, and populations can continue to increase even when juveniles or eggs suffer high mortality. Consequently, before beginning a search for antagonists of particular plant-parasitic nematodes, it is important that the vulnerable stages in the life cycle are identified and efforts directed towards locating organisms capable of attacking those stages.

During the search for better antagonists of plant-parasitic nematodes, attention should be focused on organisms with attributes likely to make them successful biological control agents. Ease of culture is obviously a prerequisite to mass production and often has been the criterion used for selecting an antagonist for introduction into soil. However, some readily cultured organisms, particularly those fungi which grow prolifically on organic substrates, are facultative parasites and can satisfy their nutritional requirements in soil without utilizing nematodes. Since their activity is not related to or synchronized with the target nematode, they may not necessarily be effective in regulating or suppressing populations of plant-parasitic nematodes.

The degree of host specificity exhibited by an antagonist is an important consideration. Polyphagous parasites and predators probably have a minimal impact on nematode populations in the natural environment and more attention should be directed towards host specific and oligophagous antagonists. Better information on the food preferences of particular antagonists is required but is not easily obtained. Observations that nematodes are consumed on agar plates or in closed chambers is a useful first step, but we need to know whether antagonists show a preference for nematodes when provided with alternative sources of food. Since some antagonists show a preference for some nematodes over others, we must also aim to search for and select those with specificity for particular plant-parasitic nematodes. For example, zoosporic fungi are more likely to attack dorylaimid nematodes than tylenchids,[137,162] while populations of *Pasteuria penetrans* vary in their pathogenicity to *Meloidogyne*.[129]

2. Problems of Establishment

The difficulty of establishing an organism in an alien environment has proved one of the major obstacles facing those interested in utilizing introduced antagonists for biological control. More than 80% of the natural enemies of insects introduced into new environments failed to establish, and only a small proportion of those that did provided adequate control of the target pest.[163] The chances of success with antagonists of nematodes is likely to be even lower because the soil environment is more complex and possibly more hostile than that existing above ground. The difficulties are compounded by the fact that an antagonist introduced into soil for the control of plant-parasitic nematodes must colonize and remain active in the rhizosphere if it is to have any chance of regulating nematode populations. This is the most microbially active zone in soil and competition between its inhabitants is intense. Fungi such as *Arthrobotrys oligospora* Fresenius and *Dactylella oviparasitica* occupy the rhizosphere in preference to the general soil mass,[25,164] (Table 6) and *Verticillium chlamydosporium* has been observed in the rhizosphere after its introduction to soil,[165] but more information is needed on the capacity of other antagonists to establish and survive in this zone.

The fungistatic nature of soil must be considered when attempts are made to establish alien fungi in soil. In fact, it has been suggested that fungistatic effects in natural soils are so severe that they are likely to prevent the successful use of both nematode-trapping and endoparasitic fungi as biological control agents.[92,106] The possibility that fungistatic effects on nematophagous fungi could be reduced or circumvented has rarely been explored and is a subject worthy of further research. Pesticides and treatments such as solarization eliminate

Table 6
PERCENTAGE OF *MELOIDOGYNE*
***INCOGNITA* EGG MASSES CONTAINING**
EGGS PARASITIZED BY *DACTYLELLA*
***OVIPARASITICA* ON TOMATO PLANTS**
GROWING IN RHIZOSPHERE OR
NONRHIZOSPHERE SOIL FROM A PEACH
ORCHARD AND DILUTED WITH STERILE
SOIL

Ratio of rhizosphere or nonrhizosphere soil: sterile soil	Rhizosphere soil	Nonrhizosphere soil
1:0	85	4
1:1	74	3
1:3	35	1
1:7	21	0
0:1	0	0

From Stirling, G. R., McKenry, M. V., and Mankau, R., *Phytopathology*, 69, 808, 1979. With permission.

major groups of microorganisms and create a partial biological vacuum in soil. Although such changes may only be temporary, they may be sufficient to allow introduced fungi to germinate and proliferate.[166] Since soil fungistasis can be annulled by the addition of energy containing nutrients to soil,[90,91] it may be possible to stimulate spore germination and growth by the judicious use of soil amendments. Fungi such as *Trichoderma*, whose conidia fail to germinate in soil because they are susceptible to fungistasis, have been established in natural soils by incubating conidia in sterile moist bran for 1 to 3 days, and using the germinated spores and young mycelium as inoculum.[167] Similar techniques might prove useful for nematophagous fungi because hyphae occupying a food base appear to be less susceptible than conidia to fungistasis.

Antagonists introduced into soil in a vegetative state or as propagules with limited food reserves are always likely to be difficult to establish in soil because they must either quickly locate a suitable host or compete with other soil organisms for alternative nutrient sources. On the other hand, parasites such as *Pasteuria penetrans, Nematophthora gynophila, Catenaria auxiliaris,* and *Verticillium chlamydosporium* may establish more readily because they produce resistant resting spores which are adapted for survival in soil. *P. penetrans* spores, for example, can remain dormant in soil for considerable periods and do not germinate until they attach to a suitable host.

3. Experimental Methods

Before an organism is tested in the field as a biological control agent against nematodes, some attempt should be made to ascertain whether it is likely to be effective against the nematode being targeted. The potential of antagonists often has been determined by assessing predacious or parasitic activity on agar plates, in observation chambers suitable for rearing microarthropods, or in sterilized soil in pots, but there is little convincing evidence that antagonistic activity in such simple systems is correlated with effects in the more complex soil environment. For example, springtails consumed large numbers of nematodes when the predator and prey were exposed on the surface of a charcoal-plaster disc, but predacious activity was reduced substantially when they were covered by a mixture of soil and vermiculite (Table 7). Nematode-trapping fungi can eliminate nematodes from agar plates[168] but similar levels of activity have never been observed in soil.

Table 7

CONSUMPTION OF *PANAGRELLUS* SP. BY TWO COLLEMBOLA SPECIES ON CHARCOAL-PLASTER DISCS WHEN NEMATODES WERE EITHER EXPOSED ON THE SURFACE OR COVERED WITH VERMICULITE OR SOIL + VERMICULITE

Collembola	Number of tests	Nematodes consumed per Collembola per day	
		Average	Range
Exposed			
Entomobryoides dissimilis	9	1120	600—2427
Sinella caeca	8	194	78—386
Covered			
Entomobryoides dissimilis	1	26	—
Sinella caeca	4	30	16—41

From Gilmore, S. K., *Search Agric.*, 1, 5, 1970. With permission.

In vitro tests may be useful preliminary screens but it is essential that additional testing be carried out in natural soil, where an introduced antagonist must survive competition from other soil organisms. R. C. Cooke and colleagues' experiments on the response of the nematode-trapping fungi to organic amendments[54,169-171] and on the susceptibility of conidia of nematophagous fungi to mycostasis[92,106] provide a good example of how the problems associated with the introduction of an antagonist into soil can be tackled through a series of simple laboratory studies. Potentially useful antagonists are often tested prematurely in the field when closely monitored laboratory and greenhouse studies may yield more information. Such studies are particularly useful for determining the application rates, timing, and types of inoculum most likely to be needed in the field, and whether establishment can be improved by modifying the soil environment with pesticides, organic amendments, or other soil treatments.

Once the decision to test an organism in the field has been made, careful attention should be given to the experimental design and to the data which is to be collected. Unfortunately, many of the conclusions drawn from experiments on biological control with introduced antagonists do not withstand critical evaluation. For example, it has often been claimed that nematode control was obtained following treatment with a particular nematophagous fungus, but it is frequently impossible to separate the effects of the fungus from those of the substrate on which the fungus was grown. Treatments such as the fungus alone, the substrate alone, and autoclaved colonized substrate should be included as standard practice, particularly when an organism is being tested for the first time. The effects of such treatments may be difficult to interpret,[172] but their inclusion creates awareness of the multiplicity of factors acting in such experiments and should lead to attempts to understand the interactions involved.

The typical field experiment, in which an antagonist is applied to field plots and nematode populations, and plant growth parameters measured at the end of the experiment, is of limited value. Some attempt should be made to assess the level of antagonistic activity during the course of the experiment, so that claims that an antagonist has acted as a parasite, predator, or toxin producer can be substantiated. In situations where direct methods cannot be used, field soil can be added to observation chambers or appropriate bioassays developed in which nematodes or their eggs are added to soil, and later recovered and examined for the effects of antagonists.[25,34,68,69]

In addition to quantifying levels of parasitism and predation, some attempt should be made to monitor populations of introduced antagonists during the course of a field experiment. Populations of predacious nematodes and microarthropods can be measured relatively easily using standard methods of extracting nematodes, mites, and springtails from soil. Resting spores of *Verticillium chlamydosporium* and *Nematophthora gynophila* can be extracted and counted[173] and it may be possible to modify this method for other fungi with readily identifiable spores. A recently developed selective medium[221] is likely to prove useful for *Paecilomyces lilacinus*. The rest of the nematophagous fungi, which form the bulk of the known antagonists of nematodes and which are most frequently tested as possible biological control agents, are difficult to monitor in soil. Quantitative methods involving dilution plates are time consuming, but this deficiency can be overcome to some extent by using a syringe technique in which the actual weight of soil placed in each plate is calculated.[174] Also, useful ecological information can be obtained by processing numerous replicated samples using one of the nonquantitative methods of isolating these fungi from soil.[175-179]

4.Production, Standardization, Formulation, Storage, Application, and Safety problems

Most of the work with antagonists of nematodes currently involves the search for potentially useful organisms and the assessment of their suitability and efficacy as biological control agents. Some of the nematode-trapping fungi and *Paecilomyces lilacinus* are being used commercially in a few countries, but the quantities being distributed are minimal and are unlikely to grow until more data on their efficacy is available. Organisms such as *Pasteuria penetrans* and *Nematophthora gynophila* have definite potential as biological control agents but are relatively host-specific and cannot be readily cultured. Consequently, research on these organisms is being concentrated on the development of in vitro culture techniques. We have not reached the stage where the problems of mass production, standardization, formulation, storage, application, and safety have been considered in detail, but these areas will become major areas of research once suitable antagonists become available.

Modern fermentation technology is widely used in the brewing industry and in the production of antibiotics, microbial insecticides, and biological herbicides[180-183] and should be amenable to the mass production of antagonists of nematodes. However, large-scale industrial production will require the development of inexpensive alternatives to the complex media currently used to culture some organisms in the laboratory. Because of variation in the concentration and viability of organisms in various preparations of a microbial product, it is imperative that differences in potency between preparations are measured. Products of constant potency can then be produced and their potency compared with that of other products. Stirling and Wachtel[126] developed a method of comparing the potency of different preparations of *Pasteuria penetrans* by modifying bioassay methods used for assaying microbial insecticides,[184] but there have been no other attempts to develop procedures for standardizing the potency of other antagonists.

Formulation of a microbial product in forms which have extended shelf lives and which can be applied to soil using conventional farm equipment are likely to present a major challenge to those interested in commercializing antagonists of nematodes. Organisms which cannot be dried without loss of viability are likely to be difficult to formulate and are unlikely to retain their activity in storage for more than a few months unless expensive storage conditions are employed.[185] The 18 month shelf life needed by a commercial product[186] is most likely to be achieved by endospore-forming bacteria and fungi which produce thick-walled oospores and chlamydospores, but some fungi have been stored for extended periods as hyphae with little loss of viability.[187]

It is imperative that formulations and methods of application are devised which are economic and which can readily be delivered with conventional farm machinery. The application of antagonists to seed or planting material has been proved one of the most efficient ways of introducing organisms into soil for the control of some plant pathogens.[1] Seed treatments are very cost effective and recent reports that *Meloidogyne incognita* and *Globodera* have been controlled by dipping potato tubers in *Paecilomyces lilacinus*[113,188] are encouraging. However, the mode of action of *P. lilacinus* applied in this manner requires more detailed investigation because competition from the indigenous soil microflora prevents many seed-inoculated fungi and bacteria from colonizing the rhizosphere more than 2 cm from treated seed.[189] If an antagonist must be added directly to soil, it should be formulated in such a way that it can be applied through farm equipment designed for applying fertilizers and granular pesticides. Most antagonists of nematodes have been tested at rates of 1 to 10 tonnes/ha (Tables 2 and 4), but these application rates will have to be reduced to no more than 400 kg/ha. This might be accomplished with systems such as diatomaceous earth granules impregnated with 10% molasses, or lignite-stillage granules, as these systems have proved suitable carriers for biological control agents used against some soil-borne fungi.[190,191]

The transposition of microorganisms to plants or soil to provide protection against pests and diseases brings biological control into the purview of government health authorities and evidence of safety to the consumer and to nontarget species must be established. Although the hazards posed by antagonists of nematodes have never been evaluated, these issues will have to be addressed in the near future because some countries now have mandatory requirements for the registration of biological pesticides. There may be risks associated with the use of fungi such as *Paecilomyces lilacinus*, which causes eye infections and facial lesions in humans and infections in domestic animals,[192] particularly when individuals have suffered physiological stress or injury prior to infection. *Trichosporon beigelii* (Kuchenm. and Rabenh.) Vuill, a parasite of young females of *Heterodera glycines* Ichinohoe,[47] causes hair and skin infections in animals and man.[193,194] Fungi such as *Aspergillus*, *Verticillium*, *Fusarium*, *Pythium*, and *Phytophthora* are sometimes associated with, or are parasitic in, eggs of plant-parasitic nematodes (Table 1) and species in these genera also cause diseases of plants and man. Since there are no known examples of an antagonist of nematodes being detrimental to a nontarget species, it is to be hoped that the extensive medical, veterinary, and plant pathological tests required to ensure that risks are minimal are not so costly as to prevent or delay the desirable development of microbial alternatives to nematicides.

V. CONSERVATION AND EXPLOITATION OF RESIDENT ANTAGONISTS

A wide range of antagonists are present in most cultivated soils and the most practical way of achieving biological control of nematodes may be to adopt practices which preserve or enhance the activity of these resident antagonists.

A. Examples
1. Stimulation of Microbial Activity with Organic Amendments
Incorporation of crop residues and organic wastes into soil often improves plant growth and reduces nematode populations. Linford et al.[77] attempted to explain this effect by suggesting that organic amendments enhanced the activity of resident antagonists of nematodes. They believed that the increase in the number of soil microorganisms which occurs in amended soil caused populations of free-living nematodes to increase. The presence of these nematodes then stimulated resident nematode-trapping fungi into predacious activity and plant-parasitic nematodes were consumed along with the free-living nematodes. This hypothesis was widely accepted and for many years organic amendments were promoted as a means of stimulating the activity of nematode-trapping fungi.

During the early 1960s, R. C. Cooke and others tried to determine the nutritional requirements of the nematode-trapping fungi and understand the balance between their saprophytic and predacious phases. Cooke[169] used an agar disc technique to record predacious

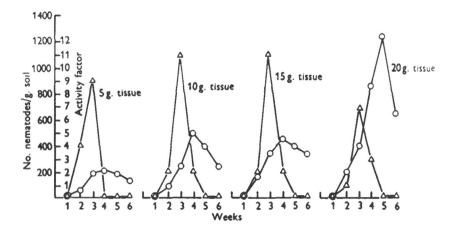

FIGURE 9.   Predacious activity of nematode-trapping, and nematode population during decomposition of different amounts of chopped cabbage leaf tissue. △, Activity factor; ○, nematode population. (From Cooke, R. C., *Trans. Br. Mycol. Soc.*, 45, 318, 1962. With permission.)

activity in a loam soil naturally infested with several nematophagous fungi. When chopped cabbage leaf tissue was added to the soil, nematode populations increased as the organic material decomposed. The presence of a large nematode population did not result in continuous predacious activity; the fungi trapped nematodes only for a short period after the onset of decomposition,[169,171] and after 6 weeks predacious activity was not detectable although nematode populations remained high (Figure 9). Similar experiments with soil amended with sucrose also showed that predacious activity was not related to the size of nematode populations.[54] Hayes and Blackburn[195] demonstrated that an energy source was a prerequisite for predacity and that the ability of nematode-trapping fungi to capture nematodes was regulated by the availability of specific nutrients. Based on the results of these studies, Cooke[54,196,197] developed the view that nematode-trapping fungi in soil are in a state of low metabolic activity under normal soil conditions and predation does not occur. When organic amendments are added to soil, nematode-trapping fungi and other zymogenous fungi are stimulated into saprophytic competition for available substrates. In the early stages of decomposition, there is intense competition for energy sources and nematode-trapping fungi, being in a unique position to use an additional substrate unavailable to other soil fungi, may be triggered into predacious activity by the nematodes present in soil. Once these energy-yielding substrates have been exhausted, predacious activity declines. Predation was seen as a means of surviving competition during periods of intense microbial activity and was not related to nematode population levels. Consequently, there are serious doubts about whether nematode-trapping fungi remain active for long enough to be totally responsible for the reduction in nematode populations which can occur following the addition of organic amendments.

It is now clear that the decomposition of organic materials in soil is a complex microbiological process in which there are marked changes in the populations of fungi and bacteria in soil. There is an initial "microbial explosion" as soluble and readily available energy sources are consumed, followed by a series of changes in the composition of the soil microflora as a succession of different microorganisms utilize the remaining chemical constituents.[198,199] It may be more than a coincidence that fungi such as *Humicola grisea* Traaen, *Gliocladium roseum*, *Phoma* sp., *Fusarium oxysporum*, *Chaetomium* sp., and *Cylindrocarpon destructans* (Zins.) Scholten, which are active in the process of cellulose decomposition[200] also are associated with or parasitic in females, cysts, and eggs of cyst nematodes (Table 1), and eggs of root-knot nematodes.[52] Perhaps parasitism by these organisms is at least partly responsible for the reduction in nematode populations which follows the addition of organic amendments. It is now recognized that a wide range of chemicals are produced during the surge in microbial activity which occurs following the addition of organic matter to soil. Ammonia, nitrites, tannins, phenols, volatile compounds, fatty acids, and other organic acids are released during the decomposition process and all these substances are either nematicidal, affect egg hatch and the motility of juveniles, or increase the resistance of roots to nematodes.[94-101,201-203] These by-products of microbial decomposition are probably largely responsible for the detrimental effects of organic amendments against nematodes, particularly when amendments have a narrow carbon:nitrogen ratio or are high in phenols and tannins.

2. Manipulation of the Environment to Favor Antagonists

Although biological control can be achieved by drastically changing the soil environment with organic amendments, it might also be accomplished through subtle changes in the environment which favor antagonists or increase the susceptibility of nematodes to antagonism. There are no examples of such strategies having been used against nematodes, but numerous possibilities exist. Rainfall during periods when adult females of *Heterodera avenae* were exposed on roots influenced the rate of infection of nematodes by *Nematophthora gynophila*,[16] and high levels of infection were observed consistently during a summer when rainfall was supplemented by overhead irrigation (Figure 10). In this situation, nematode control might be improved by using cultural practices which tend to increase soil moisture levels. Pathogenesis by *Dactylella oviparasitica* and *Pasteuria penetrans* is affected by temperature[26,204] and their effectiveness could possibly be improved by altering soil temperature regimes. Jaffee and Zehr[205] found that parasitism of *Criconemella xenoplax* by *Hirsutella rhossiliensis* was determined in part by salt concentration and suggested that parasitism might be increased by application of certain fertilizers at certain times.

Large numbers of predacious microarthropods are present in most soils,[57,59] and it may be possible to modify the soil environment so that high populations are maintained and their predacious activity enhanced. Practices such as the addition of organic matter, the maintenance of a litter layer on the soil surface, and the addition of materials which increase soil porosity and moisture retention are likely to provide a favorable environment for predacious insects and mites. However, in-depth studies will be needed to determine whether the use of such strategies results in an increase in predacious activity against plant-parasitic nematodes.

B. Aspects Requiring Further Attention

1. The Role of Amendments

Organic amendments are widely used by farmers in some countries and there is little doubt that they are beneficial to plants and detrimental to nematodes.[206] Their mode of action undoubtedly fits within the realm of biological control because soil organisms are intimately involved in their effects, either through direct parasitic, predacious, or other antagonistic activities or indirectly through the production of specific metabolic by-products or the release of nematoxic compounds from the amendment itself. However, for too long we have been content to assume that organic amendments generally act by increasing parasites and predators of nematodes in soil. There have been few attempts to provide evidence to support such claims and little in-depth research on the effects of decomposing organic materials on nematodes. If we are ever to utilize organic amendments more widely and reduce application rates to practical levels, these deficiencies must be corrected.

Organic amendments frequently are added to soil at more than 1% w/w (approximately 20 t/ha incorporated to a depth of 15 cm) and at these application rates their effects are due mainly to the action of chemicals derived from the organic additive during its decomposition. The nitrogen content of an amendment is particularly important in determining its effec-

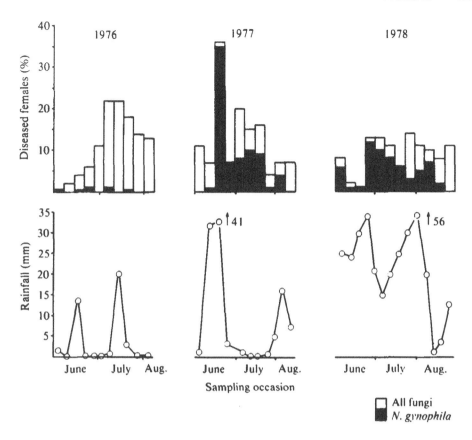

FIGURE 10. Fungal parasitism of *Heterodera avenae* females and summer rainfall in 3 years at Woburn, England. In 1978, rainfall was supplemented by overhead irrigation. (From Kerry, B. R., Crump, D. H., and Mullen, L. A., *Ann. Appl. Biol.*, 100, 495, 1982. With permission.)

tiveness against nematodes (Figure 11), and ammonia and possibly nitrite are the principle compounds thought to be responsible for this effect.[203,207-210] Tannins and phenolic compounds also play a role in the activity of other types of amendment.[99,211] The nematicidal effect of large quantities of organic matter can be utilized in situations where it is cheap and readily available or where high value crops are involved, but it is impractical or too expensive to use in broad-scale agriculture.[209] The challenge facing those interested in biological control is to develop methods of utilizing amendments at application rates which are practical for broad-scale agriculture. This might be achieved by using them to develop a microflora whose antagonistic activity augments the effects of chemicals produced during the decomposition process.

Recent work by Rodriguez-Kabana and colleagues suggests that such strategies may be successful. When chitinous waste from the seafood processing industry was incorporated into soil an adaptive microflora consisting largely of chitinolytic fungi was developed.[210,212,213] Most of these fungi were capable of parasitizing eggs of *Meloidogyne arenaria* and *Heterodera glycines* and degrading cysts of *Heterodera* and *Globodera*, and it was suggested that the reduction in plant-parasitic nematode numbers observed in response to chitin amendments may at least have been partially caused by their presence and activity in treated soils.[214]

The further development of such strategies will require detailed studies of the complex changes which occur in soil when it is amended with organic materials. It will be particularly important to know how the structure of the soil microflora changes following the addition

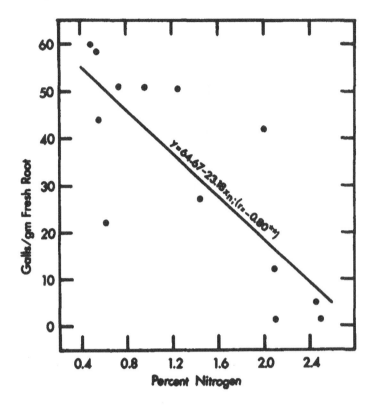

FIGURE 11. Relationship between nitrogen content of organic amendments and galling caused by *Meloidogyne arenaria* in squash plants grown in soil treated with 1% w/w of the amendments. (From Mian, I. H. and Rodriguez-Kabana, R., *Nematropica*, 12, 235, 1982. With permission.)

of the amendment and whether its composition is affected by the type of amendment or its application rate. Once this information is available, it may be possible to at least obtain-temporary nematode control by adjusting application rates and times so that peaks of antagonistic activity correspond with periods when susceptible stages of the nematode are present in the field.

VI. FUTURE PROSPECTS

Any attempt to assess the prospect of achieving biological control of nematodes in the foreseeable future is fraught with difficulties because the rate of progress will be dependent on the resources allocated for research. At present, these resources are minimal. Perusal of the plant nematological literature cited in *Helminthological Abstracts* during 1984 and 1985 showed that less than 5% of the published material was devoted to some aspect of biological control. It is estimated that world-wide no more than 20 scientists spend a major part of their time actively engaged in research on antagonists of nematodes. Biological control is in the invidious position of needing a successful example of the contrived use of an antagonist to demonstrate to administrators and funding agencies that further work is worthy of support, while lacking the resources needed to make that key advance. Without a major breakthrough, the level of inputs is unlikely to increase substantially and biological control will remain an insignificant component of nematode control programs, at least until the turn of the century. The present situation, where scientists interested in biological control are working individually in laboratories scattered across the world is another constraint on the future development of biological control. Centers specializing on biological control and staffed by scientists with skills in areas such as plant pathology, bacteriology, virology, mycology, and microbial ecology are needed. Such groups would widen the spectrum of antagonists being investigated and allow specialization in areas such as microbial ecology, where our lack of knowledge is currently inhibiting the development of biological control. Much of the work on biological control of nematodes has been superficial or descriptive and potentially useful systems rarely have been studied for long enough to fully evaluate them. Long-term, in-depth research must be encouraged.

In the past, the image of biological control has suffered because many have had unrealistic expectations of its capabilities. It has been assumed that the instant, and in some cases spectacular, results sometimes achieved with nematicides could also be obtained by simply adding an antagonist to the soil. However, it is much more likely that biological control will be slow to act and will be used as an adjunct to other methods of control. There are several examples which suggest that biological control agents and nematicides could be used together. Some fumigant nematicides have no adverse effects on nematophagous fungi,[215-217] while treatment with low concentrations of the nematicide ethoprop increases the susceptibility of nematodes to *Catenaria anguillulae*, a fungus which is not normally an aggressive parasite.[76] Spores of *Pasteuria penetrans* are not susceptible to some fumigant nematicides[34] or to heat,[218] suggesting that treatments such as preplant fumigation and solarization are likely to be compatible with the pathogen. *P. penetrans* and some nonvolatile nematicides also act synergistically,[219] perhaps because nematicides increase the motility of nematodes in soil and therefore increase the probability that they come into contact with spores of the pathogen. If biological control strategies are developed with the intent of integrating them with other nematode control practices, there is every prospect that they would extend and enhance the effectiveness of those practices.

Any attempt to consider the future of biological control leads to the question of whether the future lies with the encouragement of resident antagonists or the introduction of organisms from elsewhere. Many have considered that the introduction of an antagonist into a new environment has a limited chance of success because the alien species must be established in a situation where it faces competition, antibiosis, lysis, parasitism, and predation. However, it would be premature to dismiss this approach when so few attempts have been made to overcome the difficulties involved. Efforts should now be directed towards choosing organisms with characteristics likely to favor survival in a hostile environment, or developing techniques of inoculum preparation and environment modification which aid establishment. Because of the ubiquitous distribution of many antagonists of nematodes and the difficulties of establishing alien species, more attention is likely to be directed towards conserving and enhancing the activity of resident antagonists of nematodes. The recent work of Rodriguez-Kabana and colleagues,[212,213,214] which suggested that chitin amendments could be used to stimulate a naturally occurring, opportunistic microflora capable of antagonism towards plant-parasitic nematodes, provides cause for optimism that we may be able to manipulate the soil microflora to our advantage. However, such control strategies will not be achievable until we have a detailed understanding of the manner in which organic amendments affect the population and activity of the whole range of soil organisms which are detrimental to nematodes.

The rapidly expanding fields of biotechnology and genetic engineering are likely to have an impact on biological control, although the extent to which new technologies will influence developments is unpredictable. Nematoxic avermectins are produced in organisms such as *Streptomyces*, and it is now technically possible to transfer genes which regulate their production to bacteria which normally colonize the rhizosphere or the nematode cuticle. It may also be possible to develop new biotypes of some antagonists with properties superior to those of the wild-type strain. For example, ultraviolet radiation has been used to induce mutant strains of *Trichoderma harzianum* Rifai with resistance to certain fungicides and an enhanced ability to survive in soil. Some of these mutants were consistently more effective than the wild strain in suppressing soil-borne diseases caused by *Pythium ultimum* Trow. and *Sclerotium cepivorum* Berk.[220]

Much of the research on biological control of nematodes has tended to be oriented towards antagonists rather then towards the ultimate objective of achieving disease control, and this may have inhibited the development of effective systems of biological control. There have been many interesting studies on the biology of organisms selected largely at random from the vast array of antagonists present in soil, but many of these organisms simply have not had attributes likely to make them successful biological control agents. Progress may have been more rapid if efforts had been concentrated in areas where there were indications of antagonistic activity in the field. Fortunately, there are an increasing number of examples where some degree of biological control has been achieved through environmental manipulation or through the action of introduced or naturally occurring antagonists. We should grasp the opportunity to study these systems in detail, with the aim of understanding them, increasing their effectiveness, and extending their benefits to other areas.

REFERENCES

1. **Baker, K. F. and Cook, R. J.,** *Biological Control of Plant Pathogens,* W. H. Freeman & Co., San Francisco, Calif., 1974.
2. **Cobb, N. A.,** Transference of nematodes (mononchs) from place to place for economic purposes, *Science,* 51, 640, 1920.
3. **Linford, M. B. and Yap, F.,** Root-knot nematode injury restricted by a fungus, *Phytopathology,* 29, 596, 1939.
4. **De Bach, P.,** *Biological Control of Insect Pests and Weeds,* 1st ed., Chapman and Hall, London, 1964.
5. **Cook, R. J. and Baker, K. F.,** *The Nature and Practice of Biological Control of Plant Pathogens,* American Phytopathological Society, St. Paul, Minn., 1983.
6. **Nicholson, A. J.,** Dynamics of insect populations, *Annu. Rev. Entomol.,* 3, 107, 1958.
7. **Andrewartha, H. G. and Brich, L. C.,** *The Distribution and Abundance of Animals,* 1st ed., University of Chicago Press, Chicago, Ill., 1954, 782.
8. **Huffaker, C. B., Messenger, P. S., and De Bach, P.,** The natural enemy component in natural control and the theory of biological control, in *Biological Control,* Huffaker, C. B., Ed., Plenum Press, N.Y., 1971, chap. 2.
9. **Varley, G. C. and Gradwell, G. R.,** Recent advances in insect population dynamics, *Annu. Rev. Entomol.,* 15, 1, 1970.
10. **Knipling, E. F.,** *The Basic Principles of Insect Population Suppression and Management,* Agriculture Handbook Number 512, U.S. Department of Agriculture, U.S. Government Printing Service, Washington, D.C., 1979.
11. **Gair, R., Mathias, P. L., and Harvey, P. N.,** Studies of cereal nematode populations and cereal yields under continuous or intensive culture, *Ann. Appl. Biol.,* 63, 503, 1969.
12. **Jakobsen, J.,** The importance of monocultures of various host plants for the population density of *Heterodera avenae, Tidsskr. Planteavl,* 78, 697, 1974.
13. **Ohnesorge, B., Friedel, J., and Oesterlin, U.,** Investigations on the distribution pattern of *Heterodera avenae* and its changes in a field under continuous cereal cultivation, *Z. Pflanzenkr. Pflanzenschutz,* 81, 356, 1974.
14. **Kerry, B. R.,** Progress in the use of biological agents for control of nematodes, in *Biological Control in Crop Production,* Papavizas, G. C., Ed., Allanheld, Totowa, N.J., 1981, chap. 5.
15. **Williams, T. D.,** The effects of formalin, nabam, irrigation and nitrogen on *Heterodera avenae* Woll., *Ophiobolus graminis* Sacc. and the growth of spring wheat, *Ann. Appl. Biol.,* 64, 325, 1969.
16. **Kerry, B. R., Crump, D. H., and Mullen, L. A.,** Parasitic fungi, soil moisture and multiplication of the cereal cyst nematode, *Heterodera avenae, Nematologica,* 26, 57, 1980.
17. **Kerry, B. R.,** Fungi and the decrease of cereal cyst-nematode populations in cereal monoculture, *EPPO Bull.,* 5, 353, 1975.

18. **Kerry, B. R. and Crump, D. H.,** Observations of fungal parasites of females and eggs of the cereal cyst-nematode, *Heterodera avenae,* and other cyst nematodes, *Nematologica,* 23, 193, 1977.
19. **Kerry, B. R.,** Biocontrol: fungal parasites of female cyst nematodes, *J. Nematol.,* 12, 253, 1980.
20. **Kerry, B. R., Crump, D. H., and Mullen, L. A.,** Studies of the cereal cyst-nematode, *Heterodera avenae* under continuous cereals, 1975-1978. II. Fungal parasitism of nematode females and eggs, *Ann. Appl. Biol.,* 100, 489, 1982.
21. **Ferris, H., McKenry, M. V., and McKinney, H. E.,** Spatial distribution of nematodes in peach orchards, *Plant Dis. Rep.,* 60, 18, 1976.
22. **Stirling, G. R. and Mankau, R.,** *Dactylella oviparasitica,* a new fungal parasite of *Meloidogyne* eggs, *Mycologia,* 70, 774, 1978.
23. **Stirling, G. R. and Mankau, R.,** Parasitism of *Meloidogyne* eggs by a new fungal parasite, *J. Nematol.,* 10, 236, 1978.
24. **Stirling, G. R. and Mankau, R.,** Mode of parasitism of *Meloidogyne* and other nematode eggs by *Dactylella oviparasitica, J. Nematol.,* 11, 282, 1979.
25. **Stirling, G. R., McKenry, M. V., and Mankau, R.,** Biological control of root-knot nematodes (Meloidogyne spp.) on peach, *Phytopathology,* 69, 806, 1979.
26. **Stirling, G. R.,** Effect of temperature on parasitism of *Meloidogyne incognita* eggs by *Dactylella oviparasitica, Nematologica,* 25, 104, 1979.
27. **Jaffee, B. A. and Zehr, E. I.,** Parasitism of the nematode *Criconemella xenoplax* by the fungus *Hirsutella rhossiliensis, Phytopathology,* 72, 1378, 1982.
28. **Eayre, C. G., Jaffee, B. A., and Zehr, E. I.,** Suppression of *Criconemella xenoplax* by the fungus *Hirsutella rhossiliensis, Phytopathology,* 73, 500, 1983.
29. **Zehr, E. I.,** Evaluation of parasites and predators of plant parasitic nematodes, *J. Agric. Entomol.,* 2, 130, 1985.
30. **Sayre, R. M. and Starr, M. P.,** *Pasteuria penetrans* (ex Thorne, 1940) nom. rev., comb. n., sp.n., a mycelial and endospore-forming bacterium parasitic in plant-parasitic nematodes, *Proc. Helminthol. Soc. Wash.,* 52, 149, 1985.
31. **Williams, J. R.,** Studies on the nematode soil fauna of sugar cane fields in Mauritius. V. Notes upon a parasite of root-knot nematodes, *Nematologica,* 5, 37, 1960.
32. **Stirling, G. R. and White, A. M.,** Distribution of a parasite of root-knot nematodes in South Australian vineyards, *Plant Dis.,* 66, 52, 1982.
33. **Spaull, V. W.,** Observations on *Bacillus penetrans* infecting *Meloidogyne* in sugar cane fields in South Africa, *Rev. Nematol.,* 7, 277, 1984.
34. **Stirling, G. R.,** Biological control of *Meloidogyne javanica* with *Bacillus penetrans, Phytopathology,* 74, 55, 1984.
35. **Bursnall, L. A. and Tribe, H. T.,** Fungal parasitism in cysts of *Heterodera.* II. Egg parasites of *H. schachtii, Trans. Br. Mycol. Soc.,* 62, 595, 1974.
36. **Willcox, J. and Tribe, H. T.,** Fungal parasitism in cysts of *Heterodera* I. Preliminary investigations, *Trans. Br. Mycol. Soc.,* 62, 585, 1974.
37. **Tribe, H. T.,** Extent of disease in populations of *Heterodera,* with especial reference to *H. schachtii, Ann. Appl. Biol.,* 92, 61, 1979.
38. **Goswami, B. K. and Rumpenhorst, H. J.,** Association of an unknown fungus with potato cyst nematodes, *Globodera rostochiensis* and *G. pallida, Nematologica,* 21, 251, 1978.
39. **Nigh, E. A., Thomason, I. J., and Van Gundy, S. D.,** Identification and distribution of fungal parasites of *Heterodera schachtii* eggs in California, *Phytopathology,* 70, 884, 1980.
40. **Morgan-Jones, G. and Rodriguez-Kabana, R.,** Fungi associated with cysts of *Heterodera glycines* in an Alabama soil, *Nematropica,* 11, 69, 1981.
41. **Morgan-Jones, G., Gintis, B. O., and Rodriguez-Kabana, R.,** Fungal colonization of *Heterodera glycines* cysts in Arkansas, Florida, Mississippi and Missouri soils, *Nematropica,* 11, 155, 1981.
42. **Gintis, B. O., Morgan-Jones, G., and Rodriguez-Kabana, R.,** Mycoflora of young cysts of *Heterodera glycines* in North Carolina soils, *Nematropica,* 12, 295, 1982.
43. **Kerry, B. R., Crump, D. H., and Mullen, L. A.,** Natural control of the cereal cyst nematode, *Heterodera avenae* Woll., by soil fungi at three sites, *Crop Prot.,* 1, 99, 1982.
44. **Clovis, C. J. and Nolan, R. A.,** Fungi associated with cysts, eggs and juveniles of the golden nematode *(Globodera rostochiensis)* in Newfoundland, *Nematologica,* 29, 346, 1983.
45. **Vovlas, N. and Frisullo, S.,** *Cylindrocarpon destructans* as a parasite of *Heterodera mediterranea* eggs, *Nematol. Mediterr.,* 11, 193, 1983.
46. **Stirling, G. R. and Kerry, B. R.,** Antagonists of the cereal cyst nematode, *Heterodera avenae* Woll. in Australian soils, *Aust. J. Exp. Agric. Anim. Husb.,* 23, 318, 1983.
47. **Gintis, B. O., Morgan-Jones, G., and Rodriguez-Kabana, R.,** Fungi associated with several developmental stages of *Heterodera glycines* from an Alabama soybean field soil, *Nematropica,* 13, 181, 1983.

48. **Morgan-Jones, G., Rodriguez-Kabana, R., and Tovar, J. G.,** Fungi associated with cysts of *Heterodera glycines* in the Cauca Valley, Columbia, *Nematropica*, 14, 173, 1984.

49. **Francl, L. J. and Dropkin, V. H.,** *Glomus fasciculatum*, a weak pathogen of *Heterodera glycines*, *J. Nematol.*, 17, 470, 1985.

50. **Morgan-Jones, G., Godoy, G., and Rodriguez-Kabana, R.,** *Verticillium chlamydosporium*, fungal parasite of *Meloidogyne arenaria* females, *Nematropica*, 11, 115, 1981.

51. **Godoy, G., Rodriguez-Kabana, R., and Morgan-Jones, G.,** Fungal parasites of *Meloidogyne arenaria* eggs in an Alabama soil. A mycological survey and greenhouse studies, *Nematropica*, 13, 201, 1983.

52. **Morgan-Jones, G., White, J. F., and Rodriguez-Kabana, R.,** Fungal parasites of *Meloidogyne incognita* in an Alabama soybean field soil, *Nematropica*, 14, 93, 1984.

53. **Jones, F. G. W. and Kempton, R. A.,** Population dynamics, population models and integrated control, in *Plant Nematology*, Southey, J. F., Ed., Ministry of Agriculture, Fisheries and Food, Agricultural Development Advisory Service Publication No. GD11, London, 1978, Chap. 18.

54. **Cooke, R. C.,** The ecology of nematode-trapping fungi in the soil, *Ann. Appl. Biol.*, 50, 507, 1962.

55. **Muller, J.,** The influence of fungal parasites on the population dynamics of *Heterodera schachtii* on oil radish, *Nematologica*, 28, 161, 1982.

56. **Karg, W.,** Verbreitung und Bedeutung von Raubmilben der Cohors Gamasina als Antagonisten von Nematoden, *Pedobiologia*, 25, 419, 1983.

57. **Christiansen, K.,** Bionomics of Collembola, *Annu. Rev. Entomol.*, 9, 147, 1964.

58. **Gilmore, S. K.,** Collembola predation on nematodes, *Search Agric.*, 1, 1, 1970.

59. **Wallwork, J. A.,** Acari, in *Soil Biology*, Burges, A. and Raw, F., Eds., Academic Press, New York, 1967, chap. 11.

60. **Hale, W. G.,** Collembola, in *Soil Biology*, Burges, A. and Raw, F., Eds., Academic Press, New York, 1967, chap. 12.

61. **Steiner, G. and Heinly, H.,** The possibility of control of *Heterodera radicicola* and other plant injurious nematodes by means of predatory nemas, especially *Mononchus papillatus*, *J. Wash. Acad. Sci.*, 12, 367, 1922.

62. **Nelmes, A. J.,** Evaluation of the feeding behaviour of *Prionchulus punctatus* (Cobb), a nematode predator, *J. Anim. Ecol.*, 43, 553, 1974.

63. **Small, R. W. and Grootaert, P.,** Observations on the predation abilities of some soil dwelling predatory nematodes, *Nematologica*, 29, 109, 1983.

64. **Cohn, E. and Mordechai, M.,** Experiments in suppressing citrus nematode populations by use of a marigold and a predacious nematode, *Nematol. Mediterr.*, 2, 43, 1974.

65. **Small, R. W.,** The effects of predatory nematodes on populations of plant parasitic nematodes in pots, *Nematologica*, 25, 94, 1979.

66. **Azmi, M. I.,** Predatory behaviour of Nematodes I. Biological control of *Helicotylenchus dihystera* through the predacious nematodes *Iotonchus monhystera*, *Indian J. Nematol.*, 13, 1, 1983.

67. **Nelmes, A. J. and McCulloch, J. S.,** Numbers of mononchid nematodes in soils sown to cereals and grasses, *Ann. Appl. Biol.*, 79, 231, 1975.

68. **Stirling, G. R.,** Techniques for detecting *Dactylella oviparasitica* and evaluating its significance in field soils, *J. Nematol.*, 11, 99, 1979.

69. **Culbreath, A. K., Rodriguez-Kabana, R., and Morgan-Jones, G.,** An agar disc method for isolation of fungi colonizing nematode eggs, *Nematropica*, 14, 145, 1984.

70. **Crump, D. H. and Kerry, B. R.,** Maturation of females of the cereal cyst-nematode on oat roots and infection by an Entomophthora-like fungus in observation chambers, *Nematologica*, 23, 398, 1977.

71. **La Mondia, J. A. and Brodie, B. B.,** An observation chamber technique for evaluating potential biocontrol agents of *Globodera rostochiensis*, *J. Nematol.*, 16, 112, 1984.

72. **Kerry, B. R., Crump, D. H., and Mullen, L. A.,** Studies of the cereal cyst-nematode *Heterodera avenae* under continuous cereals, 1974-1978. I. Plant growth and nematode multiplication, *Ann. Appl. Biol.*, 100, 477, 1982.

73. **Kerry, B. R. and Crump, D. H.,** Two fungi parasitic on females of cyst-nematodes (*Heterodera* spp.), *Trans. Br. Mycol. Soc.*, 74, 119, 1980.

74. **Godoy, G., Rodriguez-Kabana, R., and Morgan-Jones, G.,** Parasitism of eggs of *Heterodera glycines* and *Meloidogyne arenaria* by fungi isolated from cysts of *H. glycines*, *Nematropica*, 12, 111, 1982.

75. **Sayre, R. M. and Keeley, L. S.,** Factors influencing *Catenaria anguillulae* infections in a free-living and a plant-parasitic nematode, *Nematologica*, 15, 492, 1969.

76. **Roy, A. K.,** Effect of ethoprop on the parasitism of *Catenaria anguillulae* on *Meloidogyne incognita*, *Rev. Nematol.*, 5, 335, 1982.

77. **Linford, M. B., Yap, F., and Oliveira, J. M.,** Reduction of soil populations of the root-knot nematode during decomposition of organic matter, *Soil Sci.*, 45, 127, 1938.

78. **Duddington, C. L., Jones, F. G. W., and Moriarty, F.,** The effect of predacious fungus and organic matter upon the soil population of beet eelworm, *Heterodera schachtii* Schm., *Nematologica*, 1, 344, 1956.

79. **Mankau, R.,** The use of nematode-trapping fungi to control root-knot nematodes, *Nematologica,* 6, 326, 1961.
80. **Hams, A. F. and Wilkin, G. D.,** Observations on the use of predacious fungi for the control of *Heterodera* spp., *Ann. Appl. Biol.,* 49, 515, 1961.
81. **Duddington, C. L., Everard, C. O. R., and Duthoit, C. M. G.,** Effect of green-manuring and a predacious fungus on cereal root eelworm in oats, *Plant Pathol.,* 10, 108, 1961.
82. **Tarjan, A. C.,** Attempts at controlling citrus burrowing nematodes using nematode-trapping fungi, *Proc. Soil Crop Sci. Soc. Fl.* 21, 17, 1961.
83. **Soprunov, F. F.,** *Predacious Hyphomycetes and Their Application in the Control of Pathogenic Nematodes,* Israel Program for Scientific Translations, Jerusalem, 1966.
84. **Al-Hazmi, A. S., Schmitt, D. P., and Sasser, J. N.,** Population dynamics of *Meloidogyne incognita* on corn grown in soil infested with *Arthrobotrys conoides, J. Nematol.,* 14, 44, 1982.
85. **Cayrol, J. C.,** Lutte biologique contre les *Meloidogyne* au moyen d' *Arthrobotrys irregularis, Rev. Nematol.,* 6, 265, 1983.
86. **Rhoades, H. L.,** Comparison of fenamiphos and *Arthrobotrys amerospora* for controlling plant nematodes in central Florida, *Nematropica,* 15, 1, 1985.
87. **Mankau, R. and Wu, X.,** Effects of the nematode-trapping fungus, *Monacrosporium ellipsosporium* on *Meloidogyne incognita* populations in field soil, *Rev. Nematol.,* 8, 147, 1985.
88. **Cayrol, J. C., Frankowski, J. P., Laniece, A., D'Hardemare, G., and Talon, J. P.,** Contre les nematodes en champignonniere. Mise au point d'une methode de lutte biologique a l'aide d'un Hyphomycete predateur: *Arthrobotrys robusta* souche 'antipolis' (Royal 300), *Pepinieristes, Horticulteurs, Maraichers, Revue Horticole,* 1984, 23, 1978.
89. **Cayrol, J. C. and Frankowski, J. P.,** Une méthode de lutte biologique contre les némétodes á galles des racines appartenant au genre *Meloidogyne, Pepinieristes Horticulteurs, Maraichers, Revue Horticole* 193, 15, 1979.
90. **Lockwood, J. L.,** Fungistasis in soils, *Biol. Rev.,* 52, 1, 1977.
91. **Mankau, R.,** Soil fungistasis and nematophagous fungi, *Phytopathology,* 52, 611, 1962.
92. **Cooke, R. C. and Satchuthananthavale, V.,** Sensitivity to mycostasis of nematode-trapping hyphomycetes, *Trans. Br. Mycol. Soc.,* 51, 555, 1968.
93. **Johnson, L. F.,** Extraction of oat straw, flax and amended soil to detect substances toxic to root-knot nematode, *Phytopathology,* 64, 1471, 1974.
94. **Khan, A. M., Alam, M. M., and Ahmad, R.,** Mechanism of the control of plant parasitic nematodes as a result of the application of oil-cakes to the soil, *Indian J. Nematol.,* 7, 93, 1974.
95. **Sitaramaiah, K. and Singh, R. S.,** Effect of atmosphere of amended soil on larval hatch of *Meloidogyne javanica* and its subsequent parasitic capacity, *Indian J. Nematol.,* 7, 163, 1977.
96. **Sitaramaiah, K. and Singh, R. S.,** Role of fatty acids in margosa cake applied as soil amendment in the control of nematodes, *Indian J. Agric. Sci.,* 48, 266, 1978.
97. **Sitaramaiah, K. and Singh, R. S.,** Effect of organic amendment on phenolic content of soil and plant and response of *Meloidogyne javanica* and its host to related compounds, *Plant Soil,* 50, 671, 1978.
98. **Alam, M. M., Siddiqui, S. A., and Khan, A. M.,** Mechanism of control of plant-parasitic nematodes as a result of the application of organic amendments to the soil. III. Role of phenols and amino acids in host roots, *Indian J. Nematol.,* 7, 27, 1977.
99. **Alam, M. M., Khan, A. M., and Saxena, S. K.,** Mechanism of control of plant parasitic nematodes as a result of the application of organic amendments to the soil. V. Role of phenolic compounds, *Indian J. Nematol.,* 9, 136, 1979.
100. **Badra, T., Saleh, M. A., and Oteifa, B. A.,** Nematicidal activity and composition of some organic fertilizers and amendments, *Rev. Nematol.,* 2, 29, 1979.
101. **Badra, T. and Elgindi, D. M.,** The relationship between phenolic content and *Tylenchulus semipenetrans* populations in nitrogen-amended citrus plants, *Rev. Nematol.,* 2, 161, 1979.
102. **Cooke, R. C.,** Ecological characteristics of nematode-trapping Hyphomycetes. I. Preliminary studies, *Ann. Appl. Biol.,* 52, 431, 1963.
103. **Cooke, R. C.,** Ecological characteristics of nematode-trapping Hyphomycetes. II. Germination of conidia in soil, *Ann. Appl. Biol.,* 54, 375, 1964.
104. **Mankau, R.,** Utilization of parasites and predators in nematode pest management ecology, *Proc. Tall Timbers Conf. Ecol. Animal Control Habitat Management,* 4, 129, 1972.
105. **Esser, R. P.,** *Monacrosporium lysipagum* infecting egg masses of *Meloidogyne acrita, J. Nematol.,* 15, 642, 1983.
106. **Guima, A. Y. and Cooke, R. C.,** Potential of *Nematoctonus* conidia for biological control of soil-borne phytonematodes, *Soil Biol. Biochem.,* 6, 217, 1974.
107. **Jaffee, B. A. and Zehr, E. I.,** Parasitic and saprophytic abilities of the nematode-attacking fungus *Hirsutella rhossiliensis, J. Nematol.,* 17, 341, 1985.

108. **Jansson, H.-B., Jeyaprakash, A., and Zuckerman, B. M.,** Control of root-knot nematodes on tomato by the endoparasitic fungus *Meria coniospora*, *J. Nematol.*, 17, 327, 1985.

109. **Dunn, M. T., Sayre, R. M., Carrell, A., and Wergin, W. P.,** Colonization of nematode eggs by *Paecilomyces lilacinus* (Thom) Sampson as observed with scanning electron microscope, *Scanning Electron Microsc.*, 3, 1351, 1982.

110. **Morgan-Jones, G., White, J. F., and Rodriguez-Kabana, R.,** Phytonematode pathology: ultrastructural studies. II. Parasitism of *Meloidogyne arenaria* eggs and larvae by *Paecilomyces lilacinus*, *Nematropica*, 14, 57, 1984.

111. **Jatala, P., Kaltenbach, T., Bocangel, M., Devaux, A. J., and Campos, R.,** Field application of *Paecilomyces lilacinus* for controlling *Meloidogyne incognita* on potatoes, *J. Nematol.*, 12, 226, 1980.

112. **Jatala, P., Salas, R., Kaltenbach, R., and Bocangel, M.,** Multiple application and long-term effect of *Paecilomyces lilacinus* in controlling *Meloidogyne incognita* under field conditions, *J. Nematol.*, 13, 445, 1981.

113. **Davide, R. G. and Zorilla, R. A.,** Evaluation of a fungus, *Paecilomyces lilacinus* (Thom) Samson, for the biological control of the potato cyst nematode, *Globodera rostochiensis* Woll. as compared with some nematicides, *Philipp. Agric.*, 66, 397, 1983.

114. **Lay, E. C., Lara, J., Jatala, P., and Gonzalez, F.,** Preliminary evaluation of *Paecilomyces lilacinus* as biological control of root-knot nematode, *Meloidogyne incognita*, in industrial tomatoes, *Nematropica*, 12, 154, 1982.

115. **Noe, J. P. and Sasser, J. N.,** Efficacy of *Paecilomyces lilacinus* in reducing yield losses due to *Meloidogyne incognita*, *Proc. 1st Int. Con. Nematol.*, Guelph, Ontario, Canada, 1, 61, 1984.

116. **Jatala, P.,** Biological control of nematodes, in *An Advanced Treatise on Meloidogyne, Biology and Control*, Vol. 1, Sasser, J. N. and Carter, C. C., Eds., North Carolina State University Department of Plant Pathology and U.S. Agency for International Development, Raleigh, N.C., 1985, chap. 26.

117. **Dickson, D. W. and Mitchell, D. J.,** Evaluation of *Paecilomyces lilacinus* as a biocontrol agent of *Meloidogyne javanica* on tobacco, *J. Nematol.*, 17, 519, 1985.

118. **Rodriguez-Kabana, R., Morgan-Jones, G., Godoy, G., and Gintis, B. O.,** Effectiveness of species of *Gliocladium, Paecilomyces and Verticillium* for control of *Meloidogyne arenaria* in field soil, *Nematropica*, 14, 155, 1984.

119. **Mankau, R. and Prasad, N.,** Infectivity of *Bacillus penetrans* in plant-parasitic nematodes, *J. Nematol.*, 9, 40, 1977.

120. **Brown, S. M. and Smart, G. C.,** Root penetration by *Meloidogyne incognita* juveniles infected with *Bacillus penetrans*, *J. Nematol.*, 17, 123, 1985.

121. **Mankau, R.,** *Bacillus penetrans* n.comb. causing a virulent disease of plant-parasitic nematodes, *J. Invertebr. Pathol.*, 26, 333, 1975.

122. **Mankau, R.,** Prokaryote affinities of *Duboscqia penetrans* Thorne, *J. Protozool.*, 22, 31, 1975.

123. **Mankau, R. and Imbriani, J. L.,** The life cycle of an endoparasite in some tylenchid nematodes, *Nematologica*, 21, 89, 1975.

124. **Sayre, R. M. and Wergin, W. P.,** Bacterial parasite of a plant nematode: morphology and ultrastructure, *J. Bacteriol.*, 129, 1091, 1977.

125. **Imbriani, J. L. and Mankau, R.,** Ultrastructure of the nematode pathogen, *Bacillus penetrans*, *J. Invertebr. Pathol.*, 30, 337, 1977.

126. **Stirling, G. R. and Wachtel, M. F.,** Mass production of *Bacillus penetrans* for the biological control of root-knot nematodes, *Nematologica*, 26, 308, 1980.

127. **Brown, S. M., Kepner, J. L., and Smart, G. C.,** Increased crop yields following application of *Bacillus penetrans* to field plots infested with *Meloidogyne incognita*, *Soil Biol. Biochem.*, 17, 483, 1985.

128. **Stirling, G. R.,** The potential of *Pasteuria penetrans* for the biological control of root-knot nematodes, *Proc. 2nd Int. Workshop Biol. Control Nematodes*, Los Banos, Philippines, in press.

129. **Stirling, G. R.,** Host specificity of *Pasteuria penetrans* within the genus *Meloidogyne*, *Nematologica*, 31, 203, 1985.

130. **Rockett, C. L. and Woodring, J. P.,** Oribatid mites as predators of soil nematodes, *Ann. Entomol. Soc. Am.*, 59, 669, 1966.

131. **Ito, Y.,** Predation by manure-inhabiting Mesostigmatids (Acarina: Mesostigmata) on some free-living nematodes, *Appl. Entomol. Zool.*, 6, 51, 1971.

132. **Sharma, R. D.,** Studies on the plant-parasitic nematode *Tylenchorhynchus dubius*, *Meded. Landbouwhogeschool Wageningen*, 71-1, 98, 1971.

133. **Muraoka, M. and Ishibashi, N.,** Nematode-feeding mites and their behaviour, *Appl. Entomol. Zool.* 11, 1, 1976.

134. **Imbriani, J. L. and Mankau, R.,** Studies on *Lasioseius scapulatus*, a mesostigmatid mite predacious on nematodes, *J. Nematol.*, 15, 523, 1983.

135. **Inserra, R. N. and Davis, D. W.,** *Hypoaspis* nr. *aculeifer:* a mite predacious on root-knot and cyst nematodes, *J. Nematol.*, 15, 324, 1983.

136. **Yeates, G. W.**, Predation by *Mononchoides potohikus* (Nematoda: Diplogasteridae) in laboratory culture, *Nematologica*, 15, 1, 1969.

137. **Boosalis, M. G. and Mankau, R.**, Parasitism and predation of soil microorganisms, in *Ecology of Soil-Borne Plant Pathogens*, Baker, K. F. and Snyder, W. C., Eds., University of California Press, Los Angeles, 1965, 374.

138. **Grootaert, P., Jaques, A., and Small, R. W.**, Prey selection in *Butlerius* sp. (Rhabditida: Diplogasteridae), *Med. Fac. Landbouww. Rijksuniv. Gent*, 42/2, 1559, 1977.

139. **Bedding, R. A.**, Low cost *in vitro* mass production of *Neoaplectana* and *Heterorhabditis* species (Nematoda) for field control of insect pests, *Nematologica*, 27, 109, 1981.

140. **Tribe, H. T.**, A parasite of white cysts of *Heterodera*, *Catenaria auxiliaris*, *Trans. Br. Mycol. Soc.*, 69, 367, 1977.

141. **Jatala, P., Kaltenbach, R., and Bocangel, M.**, Biological control of *Meloidogyne incognita acrita* and *Globodera pallida* on potatoes, *J. Nematol.*, 11, 303, 1979.

142. **Tribe, H. T.**, Pathology of cyst-nematodes, *Biol. Rev.*, 52, 477, 1977.

143. **Morgan-Jones, G., White, J. F., and Rodriguez-Kabana, R.**, Phytonematode pathology: Ultrastructural studies. I. Parasitism of *Meloidogyne arenaria* eggs by *Verticillium chlamydosporium*, *Nematropica*, 13, 245, 1983.

144. **Morgan-Jones, G. and Rodriguez-Kabana, R.**, Phytonematode pathology: fungal modes of action. A perspective, *Nematropica*, 15, 107, 1985.

145. **Walker, J. T., Specht, C. H., and Bekker, J. F.**, Nematicidal activity to *Pratylenchus penetrans* by culture fluids from actinomycetes and bacteria, *Can. J. Microbiol.*, 12, 347, 1965.

146. **Mankau, R.**, Nematicidal activity of *Aspergillus niger* culture filtrates, *Phytopathology*, 59, 1170, 1969.

147. **Guima, A. Y. and Cooke, R. C.**, Nematoxin production by *Nematoctonus haptocladus* and *N. concurrens*, *Trans. Br. Mycol. Soc.*, 56, 89, 1971.

148. **Guima, A. Y., Hackett, A. M., and Cooke, R. C.**, Thermostable nematoxins produced by germinating conidia of some endozoic fungi, *Trans. Br. Mycol. Soc.*, 60, 49, 1973.

149. **Shukla, V. N. and Swarup, G.**, Studies on root-knot of vegetables. VI. Effect of *Sclerotium rolfsii* filtrate on *Meloidogyne incognita*, *Indian J. Nematol.*, 1, 52, 1971.

150. **Desai, M. V., Shah, H. M., and Pillai, S. N.**, Effect of *Aspergillus niger* on root-knot nematode *Meloidogyne incognita*, *Indian J. Nematol.*, 2, 210, 1972.

151. **Alam, M. M., Khan, M. W., and Saxena, S. K.**, Inhibitory effect of culture filtrates of some rhizosphore fungi of okra on the mortality and larval hatch of certain plant parasitic nematodes, *Indian J. Nematol.*, 3, 94, 1973.

152. **Sakhuja, P. K., Singh, I., and Sharma, S. K.**, Effect of some fungal filtrates on the hatching of *Meloidogyne incognita*, *Indian Phytopathol.*, 31, 376, 1978.

153. **Miller, T. W., Chaiet, L., Cole, D. J., Cole, L. J., Flor, J. E., Goegelman, R. T., Gullo, V. P., Joshua, H., Kempf, A. J., Krellwitz, W. R., Monaghan, R. L., Ormond, R. E., Wilson, K. E., Albers-Schonberg, G., and Putter, I.**, Avermectins, new family of potent anthelmintic agents: isolation and chromatographic properties, *Antimicrob. Agents Chemother.*, 15, 368, 1979.

154. **Singh, S. P., Pant, V., Khan, A. M., and Saxena, S. K.**, Inhibitory effect of culture filtrates of some rhizosphere fungi of tomato as influenced by oilcakes on the mortality and larval hatch of *Meloidogyne incognita*, *Nematol. Mediterr.*, 11, 119, 1983.

155. **Khan, T. A., Azam, M. F., and Husain, S. I.**, Effect of fungal filtrates of *Aspergillus niger* and *Rhizoctonia solani* on penetration and development of root-knot nematodes and the plant growth of tomato var. Marglobe, *Indian J. Nematol.*, 14, 106, 1984.

156. **Mani, A. and Sethi, C. L.**, Some characteristics of culture filtrate of *Fusarium solani* toxic to *Meloidogyne incognita*, *Nematropica*, 14, 121, 1984.

157. **Mani, A. and Sethi, C. L.**, Effect of culture filtrates of *Fusarium oxysporum* f.sp. *ciceri* and *Fusarium solani* on hatching and juvenile mobility of *Meloidogyne incognita*, *Nematropica*, 14, 139, 1984.

158. **Vaishnav, M. U., Patel, H. R., and Dhruj, I. U.**, Effect of culture filtrates of *Aspergillus* spp. on *Meloidogyne arenaria*, *Indian J. Nematol.*, 15, 116, 1985.

159. **Jatala, P., Franco, J., Gonzales, A., and O'Hara, C. M.**, Hatching stimulation and inhibition of *Globodera pallida* eggs by the enzymatic and exopathic toxic compounds of some biocontrol fungi, *J. Nematol.*, 17, 501, 1985.

160. **Garabedian, S. and Van Gundy, S. D.**, Use of avermectins for the control of *Meloidogyne incognita* on tomatoes, *J. Nematol.*, 15, 503, 1983.

161. **Hussey, R. S. and Roncadori, R. W.**, Vesicular-arbuscular mycorrhizae may limit nematode activity and improve plant growth, *Plant Dis.*, 66, 9, 1982.

162. **Jaffee, B. A.**, Parasitism of *Xiphinema rivesi* and *X. americanum* by zoosporic fungi, *J. Nematol.*, 18, 87, 1986.

163. **Messenger, P. S. and Van den Bosch, R.**, The adaptability of introduced biological control agents, in *Biological Control*, Huffaker, C. D., Ed., Plenum Press, N.Y., 1971, chap. 3.

164. **Peterson, E. A. and Katznelson, H.,** Studies on the relationships between nematodes and other soil microorganisms. IV. Incidence of nematode-trapping fungi in the vicinity of plant roots, *Can. J. Microbiol.,* 11, 491, 1965.

165. **Kerry, B. R., Simon, A., and Rovira, A. D.,** Observations on the introduction of *Verticillium chlamydosporium* and other parasitic fungi into soil for control of the cereal cyst-nematode *Heterodera avenae, Ann. Appl. Biol.,* 105, 509, 1984.

166. **Papavizas, G. C.,** *Trichoderma and Gliocladium:* biology, ecology and potential for biocontrol, *Annu. Rev. Phytopathol.,* 23, 23, 1985.

167. **Lewis, J. A. and Papavizas, G. C.,** A new approach to stimulate population proliferation of *Trichoderma* species and other potential biocontrol fungi introduced into natural soils, *Phytopathology,* 74, 1240, 1984.

168. **Heitz, C. E.,** Assessing the predacity of nematode-trapping fungi in vitro, *Mycologia,* 70, 1086, 1978.

169. **Cooke, R. C.,** Behaviour of nematode-trapping fungi during decomposition of organic matter in soil, *Trans. Br. Mycol. Soc.,* 45, 314, 1962.

170. **Cooke, R. C.,** The predacious activity of nematode-trapping fungi added to soil, *Ann. Appl. Biol.,* 51, 295, 1963.

171. **Cooke, R. C.,** Succession of nematophagous fungi during the decomposition of organic matter in the soil, *Nature (London),* 197, 205, 1963.

172. **Baker, R., Elad, Y., and Chet, I.,** The controlled experiment in the scientific method with special emphasis on biological control, *Phytopathology,* 74, 1019, 1984.

173. **Crump, D. H. and Kerry, B. R.,** A quantitative method for extracting resting spores of two nematode parasitic fungi, *Nematophthora gynophila* and *Verticillium chlamydosporium,* from soil, *Nematologica,* 27, 330, 1981.

174. **Rodriguez-Kabana, R.,** An improved method for assessing soil fungus-population density, *Plant Soil,* 26, 393, 1967.

175. **Duddington, C. L.,** Notes on the technique of handling predacious fungi, *Trans. Br. Mycol. Soc.,* 38, 97, 1955.

176. **Barron, G. L.,** *The Nematode-Destroying Fungi,* Canadian Biological Publications, Guelph, Ontario, Canada, 1977, chap. 5.

177. **Barron, G. L.,** Nematophagous fungi: endoparasites of *Rhabditis terricola, Microb. Ecol.,* 4, 157, 1978.

178. **Mankau, R.,** A semiquantitative method for enumerating and observing parasites and predators of soil nematodes, *J. Nematol.,* 7, 119, 1975.

179. **Gray, N. F.,** Ecology of nematophagous fungi: methods of collection, isolation and maintenance of predatory and endoparasitic fungi, *Mycopathologia,* 86, 143, 1984.

180. **Smith, J. E., Berry, D. R., and Kristiansen, B.,** *Fungal Biotechnology,* Academic Press, London, 1980.

181. **Dulmage, H. T., and Rhodes, R. A.,** Production of pathogens in artificial media, in *Microbial Control of Insects and Mites,* Burges, H. D. and Hussey, N. W., Eds., Academic Press, New York, 1971, chap. 24.

182. **Soper, R. S. and Ward, M. G.,** Production, formulation and application of fungi for insect control, in *Biological Control in Crop Production,* Papavisas, G. C., Ed., Granada, London, 1981, chap. 12.

183. **Churchill, B. W.,** Mass production of microorganisms for biological control, in *Biological Control of Weeds with Plant Pathogens,* Charudattan, R. and Walker, H. L., Eds., John Wiley & Sons, New York, 1982, chap. 9.

184. **Burges, H. D. and Thomson, M. E.,** Standardization and assay of microbial insecticides, in *Microbial Control of Insects and Mites,* Burges, H. D. and Hussey, N. W., Eds., Academic Press, New York, 1971, chap. 27.

185. **Kerry, B. R.,** Biological control, in *Principles and Practice of Nematode Control in Crops,* Brown, R. H. and Kerry, B. R., Eds., Academic Press, New York, 1987, chap. 7.

186. **Couch, T. L. and Ignoffo, C. M.,** Formulation of insect pathogens, in *Microbial Control of Pests and Plant Diseases 1970-1980,* Burges, H. D., Ed., Academic Press, New York, 1981, chap. 34.

187. **Papavizas, G. C., Dunn, M. T., Lewis, J. A., and Beagle-Ristaino, J.,** Liquid fermentation technology for experimental production of biocontrol fungi, *Phytopathology,* 74, 1171, 1984.

188. **Canto-Saenz, M. and Kaltenbach, R.,** Effect of some fungicides on *Paecilomyces lilacinus* (Thom.) Samson, *Proc. 1st Int. Con. Nematol., Guelph, Canada,* 1, 13, 1984.

189. **Chao, W. L., Nelson, E. B., Harman, G. E., and Hoch, H. C.,** Colonization of the rhizosphere by biological control agents applied to seeds, *Phytopathology,* 76, 60, 1986.

190. **Backman, P. A. and Rodriguez-Kabana, R.,** A system for the growth and delivery of biological control agents to the soil, *Phytopathology,* 65, 819, 1975.

191. **Jones, R. W., Pettit, R. E., and Taber, R. A.,** Lignite and stillage: carrier and substrate for application of fungal biocontrol agents to soil, *Phytopathology,* 74, 1167, 1984.

192. **Chandler, F. W., Kaplan, W., and Ajello, L.,** *A Colour Atlas and Textbook of the Histopathology of Mycotic Diseases,* Wolfe Medical, London, 1980, 102.

193. **Gentles, J. C. and La Touche, C. J.,** Yeasts as human and animal pathogens, in *Yeasts,* Vol. 1, Rose, A. and Harrison, J., Eds., Academic Press, New York, 1969, chap. 4.

194. **Lodder, J.,** *The Yeasts, A Taxonomic Study,* 2nd ed., North Holland, Amsterdam, 1970, 1324.

195. **Hayes, W. A. and Blackburn, F.,** Studies on the nutrition of *Arthrobotrys oligospora* Fres. and *A. robusta* Dudd. II. The predacious phase, *Ann. Appl. Biol.,* 58, 51, 1966.

196. **Cooke, R. C.,** Relationships between nematode-destroying fungi and soil-borne phytonematodes, *Phytopathology,* 58, 909, 1968.

197. **Cooke, R. C.,** *The Biology of Symbiotic Fungi,* John Wiley & Sons, New York, 1977, 32.

198. **Garrett, S. D.,** *Soil Fungi and Soil Fertility,* Pergamon Press, Oxford, 1963, chap. 7.

199. **Clark, F. E.,** Bacteria in soil, in *Soil Biology,* Burges, A. and Raw, F., Eds., Academic Press, London, 1967, chap. 2.

200. **Park, D.,** Carbon and nitrogen levels as factors influencing fungal decomposers, in *The Role of Terrestrial and Aquatic Organisms in Decomposition Processes,* Anderson, J. M. and Macfayden, A., Eds., Blackwell Scientific, Oxford, 1976, chap. 3.

201. **Rodriguez-Kabana, R., Jordan, J. W., and Hollis, J. P.,** Nematodes: biological control in rice fields: role of hydrogen sulphide, *Science,* 148, 524, 1965.

202. **Hollis, J. P. and Rodriguez-Kabana, R.,** Rapid kill of nematodes in flooded soil, *Phytopathology,* 56, 1015, 1966.

203. **Mian, I. H. and Rodriguez-Kabana, R.,** Survey of the nematicidal properties of some organic materials available in Alabama as amendments to soil for control of *Meloidogyne arenaria, Nematropica,* 12, 235, 1982.

204. **Stirling G. R.,** Effect of temperature on infection of *Meloidogyne javanica* by *Bacillus penetrans, Nematologica,* 27, 458, 1981.

205. **Jaffee, B. A. and Zehr, E. I.,** Effects of certain solutes, osmotic potential and soil solutions on parasitism of *Criconemella xenoplax* by *Hirsutella rhossiliensis, Phytopathology,* 73, 544, 1983.

206. **Muller, R. and Gooch, P. S.,** Organic amendments in nematode control. An examination of the literature, *Nematropica,* 12, 319, 1982.

207. **Heubner, R. A., Rodriguez-Kabana, R., and Patterson, R. M.,** Hemicellulosic waste and urea for control of plant-parasitic nematodes: effect on soil enzyme activities, *Nematropica,* 13, 37, 1983.

208. **Walker, J. T.,** Populations of *Pratylenchus penetrans* relative to decomposing nitrogenous soil amendments, *J. Nematol.,* 3, 43, 1971.

209. **Mian, I. H. and Rodriguez-Kabana, R.,** Soil amendments with oil cakes and chicken litter for control of *Meloidogyne arenaria, Nematropica,* 12, 205, 1982.

210. **Mian, I. H., Godoy, G., Shelby, R. A., Rodriguez-Kabana, R., and Morgan-Jones, G.,** Chitin amendments for control of *Meloidogyne arenaria* in infested soil, *Nematropica,* 12, 71, 1982.

211. **Mian, I. H. and Rodriguez-Kabana, R.,** Organic amendments with high tannin and phenolic contents for control of *Meloidogyne arenaria* in infested soil, *Nematropica,* 12, 221, 1982.

212. **Godoy, G. R., Rodriguez-Kabana, R., Shelby, R. A., and Morgan-Jones, G.,** Chitin amendments for control of *Meloidogyne arenaria* in infested soil. II. Effects on microbial population, *Nematropica,* 13, 63, 1983.

213. **Rodriguez-Kabana, R., Godoy, G., Morgan-Jones, G., and Shelby, R. A.,** The determination of soil chitinase activity: conditions for assay and soil ecological studies, *Plant Soil,* 75, 95, 1983.

214. **Rodriguez-Kabana, R., Morgan-Jones, G., and Gintis, B. O.,** Effects of chitin amendments to soil on *Heterodera glycines,* microbial populations and colonization of cysts by fungi, *Nematropica,* 14, 10, 1984.

215. **Mankau, R.,** Effect of nematocides on nematode-trapping fungi associated with the citrus nematode, *Plant Dis. Rep.,* 52, 851, 1968.

216. **Mankau, R. and Imbriani, J.,** Tolerance and uptake of 1,2-dibromoethane by nematode-trapping fungi, *Phytopathology,* 61, 1492, 1971.

217. **Mitsui, Y.,** Fungicidal effect of nematicides on nematode-trapping fungi, *Jpn. J. Nematol.,* 1, 25, 1972.

218. **Dutky, E. M. and Sayre, R. M.,** Some factors affecting infection of nematodes by the bacterial spore parasite *Bacillus penetrans, J. Nematol.,* 10, 285, 1978.

219. **Brown, S. M. and Nordmeyer, D.,** Synergistic reduction in root galling by *Meloidogyne javanica* with *Pasteuria penetrans* and nematicides, *Rev. Nematol.,* 8, 285, 1985.

220. **Papavizas, G. C., Lewis, J. A., and Abd-El Moity, T. H.,** Evaluation of new biotypes of *Trichoderma harzianum* for tolerance to benomyl and enhanced biocontrol capabilities, *Phytopathology,* 72, 126, 1982.

221. **Mitchell, D. J., Kannwischer-Mitchell, M. E., and Dickson, D.,** A semi-selective medium for the isolation of *Paecilomyces filacinus* from soil, *J. Nematol.,* 19, 255, 1987.

INDEX

INDEX

A

Acari (mites), 79—81

N-Acetylgalactosamine, in inhibition of nematode
capture, 64

Acremonium bacillisporum, colonization of *Ascaris*
eggs, 46

Acremonium strictum, colonizaiton of *Heterodera*
schachtii cysts/eggs, 44

Acrobeloides nanus, consumption by Insecta, 77

Acrostralagmus, taxonomy, 15

Acrostralagmus obovatus, 14, 15, 25

Adenoplea, nematodes attacked by, 75

Adhesion, see *Acrobotrys oligospora*, adhesion
process

Adhesive, see Predatory nematophagous fungi

Age, vulnerability of nematode eggs to fungal
colonization and, 49

Allolobophora caliginosa, 76

Alternaria, colonization of *Heterodera*, 44

Amendments (organic) to soil, 77, 109—113, 126—
130

Amino acids, trap formation and, 24

Antagonists, see Biocontrol of nematodes

Antibiotics, from nematophagous fungi, 23

Aphelenchoides, fungi parasitizing, 21

Aphelenchoides ritzemabosi, consumption by *Eisenia*
foetida, 76

Aphelenchoides rutgersi, protozoa attacking, 75

Aphelenchs, predatory, 82

Aphelenchus avenae, protozoa attacking, 75

Aquatides thornei, 84

Arid ecosystems, litter breakdown in, mite-nematode
interactions, 81

Arthrobotrys, conidia of, 15

Arthrobotrys amerospora, in biocontrol, 113

Arthrobotrys arthrobotryoides, 23, 61, 69, 110

Arthrobotrys botryospora, nematode trapping by, 8

Arthtrobotrys brochopaga, in biocontrol, 110

Arthrobotrys candida, in biocontrol, 110, 111

Arthrobotrys conoides, 22, 66, 69, 110, 112

Arthrobotrys constringens, habitat, 25

Arthrobotrys dactyloides, 105

    in biocontrol, 110

    conidia of, 15

    constricting rings, 12

    horizontal distribution in soil, 32

    nematicidal activity, 68, 69

Arthrobotrys dolioformis, in biocontrol, 112

Arthrobotrys ellipsospora, 22, 27

Arthrobotrys irregularis, 29, 32

Arthrobotrys javanica, conidia of, 15

Arthrobotrys kirkhizica, in biocontrol, 112

Arthrobotrys musiformis, 15, 25, 110, 111

Arthrobotrys oligospora, 4, 25, 122

    adhesion process, 64—66

    adhesive net, 9, 10, 61

attraction intensity, 23

attraction of nematodes, 61

in biocontrol, 110

carbohydrate specialization of trap lectins, 22

conidia of, 8

growth rate, herbicide effects on, 30

horizontal distribution in soil, 32

host range, 22

lectin-binding carbohydrate for, 66

lectin-carbohydrate binding, 22

nematicidal toxin, 68, 69

in nematode cysts, 41

nematode trapping by, 9—10

nutrition, 20

penetration of *Panagrellus redivivus*, 67—68

trap formation in, 24

Arthrobotrys oviformis, trap formation in, 24

Arthrobotrys pauca, 12, 25

Arthrobotrys pectospora, vertical distribution in soil,
31

Arthrobotrys pravicovii, in biocontrol, 112

Arthrobotrys robusta, 25, 29, 109, 111

Arthrobotrys superba, 9, 14, 15, 109, 112

Arthrobotrys thaumasia, in biocontrol, 110, 111

Ascaris, fungal colonization of, 46, 54

Aureobasidiun pullulans, colonization of *Meloido-*
gyne incognita eggs, 46

B

Bacteria, in biocontrol of nematodes, 118—120

Baermann funnel method, 18

Basidiomycetes, 5, 6

Belonolaimus, 77

Biocontrol of nematodes, 74, 87—88, 93—132

    Adenoplea in, 75

    classic approach to, 108—120

    conservation and exploitation of resident
        antagonists, 126—130

    definition, 94

    fungistatic effects of soil and, 122—123

    future prospects, 130—132

    introduction of antagonists, 108—126, 131

        bacteria, 118—120

        egg-parasitic fungi, 115—118

        endoparasitic fungi, 114—115

        experimental methods for assessment of
            effectiveness of antagonist, 123—125

        host specificity of, 122

        life cycle of nematode most vulnerable to, 121—
            122

        mass production of antagonists, 122

    nematode-trapping fungi, 109—114

        predatory nematodes and microarthropods, 120

        problems of establishment of antagonists, 122—
            123

        production, standardization, formulation,storage,

Printed and bound by CPI Group (UK) Ltd, Croydon, CR0 4YY

22/10/2024

01777632-0014